Many Faces Many Microbes

Personal Reflections in Microbiology

Many Faces Many Microbes

Personal Reflections in Microbiology

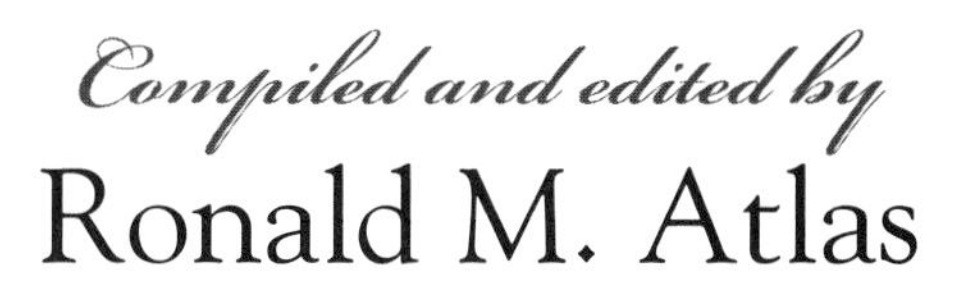

Ronald M. Atlas

ASM Press
Washington, D.C.

ASM Press
American Society for Microbiology
1752 N Street, N.W.
Washington, DC 20036-2804

Library of Congress Cataloging-in-Publication Data

Many faces, many microbes / compiled and edited by Ronald M. Atlas.
p. cm.
Includes bibliographical references.
ISBN 1-55581-190-6
1. Microbiologists--Biography. I. Atlas, Ronald M., 1946-

QR30 .M36 2000
579'.092'2--dc21
[B]

99-098184

Printed in the United States of America

Contents

Contributors

Julius Adler
Department of Biochemistry, University of Wisconsin Medical School, 433 Babcock Drive, Madison, WI 53706-1544

Ronald M. Atlas
Department of Biology, University of Louisville, Life Science Building #139, Louisville, KY 40292

Albert Balows
Department of Biology, Georgia State University; 105 Bay Colt Road, Alpharetta, GA 30004

Ellen Jo Baron
Clinical Microbiology/Virology Laboratory, Stanford University Hospital, 300 Pasteur Drive, RM H1537-J, Stanford, CA 94305-5250

Joan Wennstrom Bennett
Department of Cellular and Molecular Biology, Tulane University School of Medicine, 2000 Percival Stern Hall, New Orleans, LA 70118-5698

Kenneth Berns
Office of the Dean, University of Florida College of Medicine, P.O. Box 100014, Room H-102, Gainesville, FL 32610-0014

Jean Brenchley
Department of Microbiology and Biotechnology, Pennsylvania State University College of Medicine, 209 S. Frear, University Park, PA 16802-1009

Raúl Cano
Department of Biological Sciences, California Polytech State University, San Luis Obispo, CA 93407

A. M. Chakrabarty
Department of Microbiology, University of Illinois Medical Center, 835 S. Wolcott Avenue, Chicago, IL 60612-7340

Rita R. Colwell
National Science Foundation, 4201 Wilson Boulevard, Room 1205, Arlington, VA 22230

Julian Davies
Department of Microbiology and Immunology, University of British Columbia, Vancouver, BC, V6T1Z3, Canada

Erik D. A. De Clercq
Rega Institute, 10 Minderbroedersstraat, Leuven B-3000, Belgium

Brendlyn D. Faison
Department of Biological Sciences, Hampton University, Hampton, VA 23668

Stanley Falkow
Department of Microbiology and Immunology, Stanford University School of Medicine, 299 Campus Drive, Fairchild Room D09A, Stanford, CA 94305-5124

Claire M. Fraser
The Institute for Genomic Research, 9712 Medical Center Drive, Rockville, MD 20850

Herman Friedman
Department of Medical Microbiology and Immunology, University of South Florida College of Medicine, 12901 Bruce Downs Boulevard, Box 10, Tampa, FL 33612-4742

Susan Gottesman
Laboratory of Molecular Biology, National Cancer Institute, Building 37, Room 2E18, Bethesda, MD 20892-4255

David A. Hopwood
John Innes Centre, Norwich Research Park Colney, Norwich, Norfolk NR4 7UH, England

Jennie Hunter-Cevera
Office of the President, University of Maryland Biotechnology Institute, 4321 Hartwick Road, Suite 550, College Park, MD 20740

John Lyman Ingraham
Department of Bacteriology, University of California, Davis, School of Medicine, 7535 Cheryl Lane, Fair Oaks, CA 95628

Holger W. Jannasch
Woods Hole Oceanographic Institution, Woods Hole, MA [deceased]

Samuel Kaplan
Department of Microbiology and Molecular Genetics, University of Texas–Houston Medical School, 6431 Fannin, Houston, TX 77030

John Lennox
Department of Microbiology, Pennsylvania State University–Altoona, 152 CLRC, Altoona, PA 16603

Jay A. Levy
Division of Hematology/Oncology, University of California, San Francisco, School of Medicine, Box 1270, San Francisco, CA 94143-1270

Stuart B. Levy
Center for Adaptation Genetics and Drug Resistance, Tufts University School of Medicine, 136 Harrison Avenue, Boston, MA 02111-1800

Josephine A. Morello
Department of Pathology, University of Chicago Pritzker School of Medicine, MC0001, 5841 S. Maryland Avenue, Chicago, IL 60637-1470

Carol Nacy
Sequella, Inc., 9610 Medical Center Drive, Suite 200, Rockville, MD 20850

Cynthia A. Needham
ICAN Productions, P.O. Box 3599, Stowe, VT 05672-3599

Eugene W. Nester
Department of Microbiology, University of Washington School of Medicine, Box 357242, Seattle, WA 98195

C. J. Peters
Centers for Disease Control and Prevention, 1600 Clifton Road, MS A-26, Atlanta, GA 30333

Frederic K. Pfaender
Department of Environmental Science and Engineering, University of North Carolina at Chapel Hill, School of Public Health, South Columbia Street, Chapel Hill, NC 27599-7400

Michael A. Pfaller
Medical Microbiology Division, 606 GH Department of Pathology, University of Iowa College of Medicine, Iowa City, IA 52242

Moselio Schaechter
Department of Biology, San Diego State University, San Diego, CA 92120

David Schlessinger
Laboratory of Genetics, National Institute on Aging, National Institutes of Health, 4940 Eastern Avenue, Baltimore, MD 21224

Cathy Squires
Department of Molecular Biology and Microbiology, Tufts University School of Medicine, 136 Harrison Avenue, Boston, MA 02111-1800

Erko Stackebrandt
DSMZ-Deutsche Sammlung von Mikroorganismen und Zellkulturen GmbH, Mascheroder Weg 1b, Braunschweig 38124, Germany

Fred C. Tenover
Centers for Disease Control and Prevention, Hospital Infection Program, 1600 Clifton RD NE G08, Atlanta, GA 30333

Lucy S. Tompkins
Departments of Medicine and Microbiology, Immunology, and Pathology, Stanford University Medical Center, 300 Pasteur Drive, Room H1537-J, Stanford, CA 94305-1901

Amy Cheng Vollmer
Department of Biology, Swarthmore College, 500 College Avenue, Swarthmore, PA 19081-1397

Luther Williams
Payson Center for International Development and Technology Transfer, Tulane University, Arlington, VA 22203

Carl R. Woese
Department of Microbiology, University of Illinois at Chicago College of Medicine, 601 South Goodwin Avenue, 131 Burrill Hall, MC 110, Urbana, IL 61801

Ralph Wolfe
Department of Microbiology, University of Illinois at Chicago College of Medicine, 601 South Goodwin Avenue, B 103 Chemical and Life Sciences Laboratory, Urbana, IL 61801

Preface

When Paul de Kruif wrote *Microbe Hunters* in 1926, he hoped that it would pique interest in the mysterious, unseen world of microbes and those scientists who dared to become microbiologists. "I always did like these microbe hunters, because they were *human*, making mistakes, quarreling, trying, mostly failing but barging back into the battle again, not giving up. They weren't all brain with no body nor were they stuffed shirts with empty skulls as you sometimes find folks with big names to be. The best thing about them was that they stayed kids after they'd grown up. That's why they had such a good time digging and digging at jobs most people would call monotonous. They were really just playing." Through *Microbe Hunters*, a book that became required grade-school reading, future generations were inspired with a fascination of those invisible strangers and those who pioneered the scientific journey into the microbial world.

Many Faces—Many Microbes is a continuation of the story of the beginnings of microbiology told by De Kruif—of the quest to pique the interest of future generations in the wonderful world of microbiology. It is a compilation of a diverse collection of contemporary microbiologists' life stories—highlighting how and why each became a microbiologist and illustrating the fascinating breadth of the field. It is an attempt to demystify science by revealing the human faces of leading scientists. Today's microbiologists are truly reflective of society: they are men and women; young and, well, older; immigrants and native-born; black, white, Hispanic, and other—just like society at large, united by an interest in the acquisition and application of knowledge about microbes and how they affect our lives.

As we enter the twenty-first century, the American Society for Microbiology (ASM), the world's largest organization of biologists, celebrates its 100th anniversary. This book honors ASM and its membership by introducing a few of the wide variety of members. Two of the themes of the centennial celebration of the ASM are education and communication. It is hoped that this book will contribute to the Society's success in both of these missions: to improve science education and to improve communication between scientists and the general public. An informed, scientifically literate public is increasingly essential in our democracy as more and more science-based issues become public and political confrontations. Scientific literacy must be based on improved science education, especially in the microbiological sciences, at all levels from preschool to postgraduate.

Above all, *Many Faces—Many Microbes* is a tale to entice a new generation of microbiologists to take up the quest—to embark on the career path we chose by becoming microbiologists—to capture the excitement that drives us to explore the unseen world of those invisible strangers.

Ronald M. Atlas

FROM THE BOOK OF INSPIRATION

Just over three centuries ago, "an obscure man named van Leeuwenhoek looked for the first time into a mysterious new world peopled with a thousand different kinds of tiny beings, some ferocious and deadly, others friendly and useful, many of them more important to mankind than any continent or archipelago.... This is the story of van Leeuwenhoek, the first of the microbe hunters. It is the tale of the bold and persistent and curious explorers and fighters of death who came after him. It is the plain history of their tireless peerings into this new fantastic world. They have tried to chart it, these microbe hunters and death fighters....

"I love these microbe hunters, from old Antony van Leeuwenhoek to Paul Ehrlich. Not especially for the discoveries they have made nor for the boons they have brought mankind. No. I love them for the men they are. I say they *are*, for in my memory every man jack of them lives and will survive.... It is as sure as the sun following the dawn of tomorrow that there will be other microbe hunters to mold other magic bullets, surer, safer, bullets to wipe out for always the most malignant microbes of which this history has told...."

So concluded Paul de Kruif in his classic 1926 book *Microbe Hunters*. Bringing a human face to the early days of microbiology,

De Kruif captured the drive and inquisitive spirit of the earliest investigators of the invisible microbial world. Although his book ends with Paul Ehrlich's discoveries in the first decades of the twentieth century, this is only really where our stories begin. Here is our continuing saga of men and women in quest of these invisible and fascinating strangers.

1

STANLEY FALKOW

Toward Understanding the Molecular Basis of Bacterial Pathogenicity

When I was about eleven years old in Newport, Rhode Island, I happened on a book in the library called *Microbe Hunters* by Paul de Kruif. I do not know why I selected this volume. I actually was looking for something on astronomy, which was the center of my burgeoning interest in science. I read *Microbe Hunters* in one sitting. When I put it down, there was no doubt in my mind that I wanted to be a bacteriologist—to be precise, a medical bacteriologist. In subsequent years, I have met many people who were similarly affected by this book. Most of them were drawn to study medicine, but I was more taken by the stories of Pasteur, Koch, Ehrlich, and Metchnikoff. I wanted to study microbes, not people. I wanted to spend my life in a laboratory rather than at the bedside.

Although during my high school years I did not swerve from my decision to spend my life studying bacteria, it is fair to say that my preparation left a great deal to be desired. I was a terrible student. Yet I managed to be accepted at the University of Maine and majored in bacteriology. At Orono, I was the lucky beneficiary of the attentions of several dedicated teachers. They began my initiation into the world of microbiology. They were the first to show me that teaching is a rewarding occupation and that the best teachers learn from their students.

As my first year of college came to an end, I secured a position as a volunteer at the Newport Hospital. There I was taught aspects of medical bacteriology, serology, and hematology in exchange for my services as an assistant for postmortem examinations. This was the real stuff! Every day, after drawing preoperative blood samples, I sat next to an experienced clinical bacteriologist, Alice Sauzette, who taught me the procedures for identifying microorganisms from clinical samples. My passion became Gram staining every bit of patient material I could obtain despite its appearance, odor, or state of preservation.

Observing microbes from individuals with infectious diseases formed the roots of my thinking about bacterial pathogenicity. That summer's experience was repeated each summer thereafter during my college years and was my full-time occupation during the two years between finishing my undergraduate degree and entering graduate school. This experience permitted me to work with most of the pathogenic bacteria mentioned in contemporary textbooks. In later years, my laboratory became known for working on many different pathogenic species rather than focusing on only one or at most two pathogenic species. In part this was simply a reflection of the practical experience I gained as an apprentice in the clinical laboratory.

My professors at the University of Maine were pleased with my progress in learning practical medical microbiology in the hospital laboratory. They cautioned me not to mistake technical skill in recovering bacteria from clinical material for discovery research. To be a successful scientist, one needs a modicum of technical skill, and a good clinical microbiologist must understand the basic biology of microorganisms. As a novice scientist, I needed to be taught how to design controlled experiments and that the failed experiment was sometimes more informative than the successful one.

I was fortunate to be nurtured by so many caring people as an undergraduate. I was expected to complete an independent research project. My professor, E. R. Hitchner, guided me through work on *Proteus*, which led to my first scientific publication in 1956. He and others at the university made sure that my schooling in microbiology went beyond medical bacteriology. Their insistence that I study soil and dairy bacteriology paid enormous dividends later in my career when I was first challenged to understand the dynamic interactions of complex microbial populations and of microorganisms with more complex co-resident living forms.

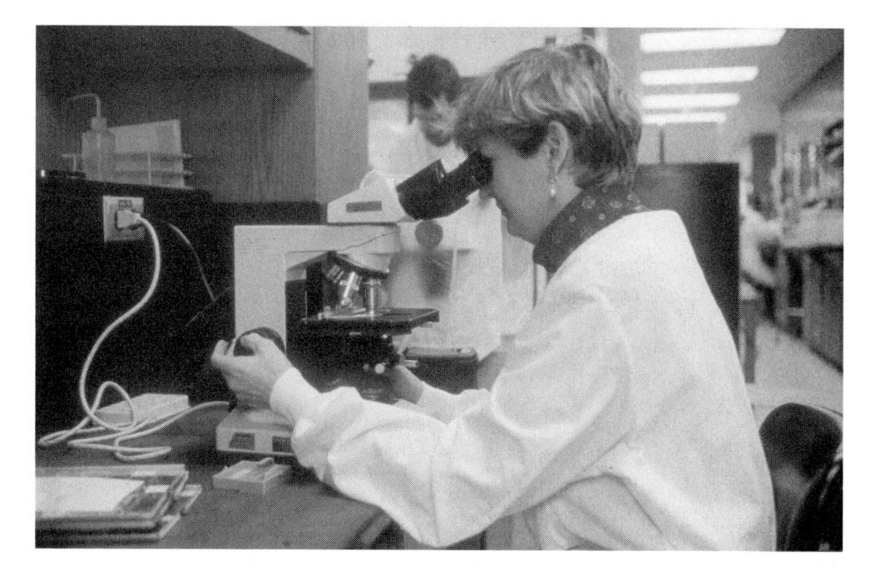

Viewing clinical specimens.

I began my graduate studies at Brown University with C. A. Stuart (the namesake of *Providencia stuartii*) at the dawn of what Nobel Prize winner Salvador Luria called the "golden age of molecular biology." Hardly a day went by without a new revelation or a new way to study the genetic and molecular basis of living things, particularly bacteria and bacterial viruses; not many people worked on pathogenic bacteria. At Brown, I began my first experiments with the goal of trying to understand the genetic basis of bacterial pathogenicity. Bacterial genetics was still a young field; isolating DNA from bacteria was not a routine procedure. Nevertheless I was able to make some progress in understanding the genetic organization of the *Salmonella* chromosome. In parallel, I began experiments on the molecular nature of bacterial plasmids (called episomes in those days), particularly the class of plasmids that encode resistance to antibiotics and that became known as the R factors.

My first job after graduate school was at the Walter Reed Army Institute of Research. It was one of the few places in the world investigating the genetics of pathogenic bacteria. From 1960 through 1966, I worked with two wonderful older investigators, Sam Formal and Louis Baron, trying to learn how *Salmonella typhi* and *Shigella flexneri* cause disease and searching for the basis for a protective vaccine. It was a wonderful time in my life because I was young enough to spend most of my waking hours doing experiments. It was also a frustrating time because the experimental tools available to us were insufficient to dissect out fundamental questions, such as why *Escherichia coli* was normally a commensal in the colon although *S. flexneri*, which was 70% identical to *E. coli* at the DNA homology level, was pathogenic.

In 1966, at the age of 32, I established my own laboratory at Georgetown University and began to pass on the legacy of my mentors, who had invested part of their lives in mine. It is one of the marvelous benefits of being a scientist. Each day for the past thirty years, I have entered my laboratory, a world populated by young people mostly between the ages of 21 and 35. The language changes ("Oh, wow" to "Cool") and the music changes (from Pete Seeger to the Dead Kennedys to rap), but the excitement of science and discovery remains as fresh for me as it was when I began my life as a bacteriologist as a freshman at the University of Maine. It is an extraordinary privilege to be able to touch someone's life in the same way my life was changed by a group of scientists who passed their knowledge on to me.

Over the years, my laboratory has progressed with the goal of gaining understanding of the genetic and molecular basis of bacterial pathogenicity. The study of pathogenicity is not restricted to an examination of diseases. A pathogen is, in fact, a specialized microorganism that has an essential requirement for growth and replication in a particular host organism (there are pathogens of amoebae and nematodes as well as those of humans). In the beginning, my students and I examined the determinants of pathogenicity that were carried on plasmids. This reflected, in part, the practical fact that smaller DNA molecules, like plasmids, were the easier targets of research, rather than the entire chromosome of a microorganism.

Thus, in the early 1970s, we began a study of *E. coli* plasmid-mediated enterotoxins associated with traveler's diarrhea. At about the same time, the methods of gene cloning were just beginning to be understood. We were the first to apply these methods to the study of bacterial pathogens. Initially, we cloned the genes for the *E. coli* enterotoxins. In parallel, I might also note, we applied some of the methods of recombinant DNA to the study of the epidemic spread of certain bacteria and used molecular methods to "fingerprint" bacteria isolated from cholera, antibiotic-resistant gonococci throughout Africa and Asia, and antibiotic-resistant bacteria in the hospital setting. Once we had the genes that encoded enterotoxins, it occurred to several of my students that these genes could serve as signals for the presence of pathogenic bacteria in patient material. Instead of attempting to isolate the living bacteria from patients, we determined by DNA hybridization that the presence of an enterotoxin gene could be a useful surrogate in large-scale epidemiologic studies. It was one of the first applications of DNA probes for the identification of pathogenic bacteria.

The quick succession of methods, such as gene cloning, DNA sequencing, the polymerase chain reaction, and other remarkable technological breakthroughs, enabled us in the 1990s to expand our investigation of bacterial pathogenicity to encompass an ever-growing list of microorganisms. My students and I have productively studied the mechanisms of pathogenicity of many diverse organisms, including *Salmonella*, *Shigella*, enterotoxigenic and uropathogenic *E. coli*, *Bordetella pertussis*, the gonococcus, *Yersinia* spp., *Haemophilus influenzae*, and, more recently, pathogenic mycobacteria, *Helicobacter pylori*, and *Legionella pneumophila*. In the beginning, we attempted to identify the genes important in pathogenicity and to clone and sequence this genetic information. As time went on, however, we began to apply the same criterion to the study of genes associated with pathogenicity as Robert Koch (one of the heroes of *Microbe Hunters*) applied to establish that certain bacteria were the causative agents of human disease. We simply made the assumption that if a gene were essential for pathogenicity, it would be present in an active form in pathogens. Thus, using molecular methods, we would mutate a gene we thought was important in pathogenicity to an inactive form to determine its effect on the organism's virulence. If virulence was reduced, we suspected that the gene was directly or indirectly essential for pathogenicity. To prove the point, we once again reintroduced an active form of the gene to restore pathogenicity.

Pathogenicity is a complex and multigenic phenotype. It is tightly regulated by microorganisms; the genes important in pathogenicity are usually used only as they are needed, and their expression is often regulated by environmental cues from the infected host animal. The genetic choreography of a pathogen in contact with its host is just beginning to be understood. Since the mid-1990s, there has been much more of an ecumenical approach to the study of bacterial pathogens. It is not enough to understand the bacterial genes and their regulation. One must understand the cell biology and cellular immunology of the host organisms as well. The way humans look and act, after all, is in part a reflection of the fact that we reside in a world teeming with microorganisms. Our survival depends on the commensal and symbiotic microorganisms with which we coexist. At the same time, our bodies are designed to restrict the entry and growth of unwanted microorganisms. Only a few microbes are pathogenic for us. These microbes have learned secrets, which they keep from other microorganisms, on how to breach our local defenses and gain entry into our body, where they can replicate sufficiently to establish themselves or be transmitted to a new susceptible host.

Recently I heard a student comment that microbiology was going to be boring in the future because all the chromosomes of bacteria would be sequenced in the near future. I suspect it is true that DNA sequencing of most of the important human pathogens will be accomplished in the first decade of the twenty-first century. It would be wonderful to think that simply sequencing DNA will provide us with a full understanding of genetic function, but to say that we will understand the biology of microorganisms (or any other creatures) from their DNA sequence alone is as silly as saying we understand humans because we know the proportion of elements that make up their protoplasm. DNA sequencing removes some of the tedium from research. It is a great technological achievement, but it does not substitute for creativity or discovery science. One of our great challenges is to find ways to follow microorganisms in the host animal—in real time. We need to design sophisticated genetic and biochemical reporting systems that allow us to examine gene expression and the order of gene expression as the biology of the pathogen unfolds during the infectious process.

Yet it is clear that technology is progressing at an amazing pace—one much faster than I could have imagined just a few years ago. We may soon be able to unravel many of the remaining mysteries of pathogenicity by using the combined power of the new field of micro-DNA arrays and genomics. It is now possible to have every known bacterial gene of a given species literally printed as a series of spots on an ordinary glass slide. One can likewise have an analogous array of genes from a human, mouse, or any other organism. So it is possible to analyze the global expression of a pathogen and the host response to the pathogen in a single experiment. Although this technology is being applied initially to model infections of tissue culture cells, by the time this book is published, experiments will be performed in infected animals and even in biopsy material.

One sobering feature of this technological breakthrough is the amount of information that can be garnered at one time. It will not be unusual to collect 250,000 data points from a single experiment. My students, for example, examined the gene expression of human epithelial cells in culture with *H. pylori*. The examination of just the top 1% of genes expressed was daunting, all the more so since three of the top five expressed genes were expressed sequence tags of unknown function.

Because of the explosive development of new tools for studying microorganisms, I am more excited to be a microbiologist—being able to continue the process of discovery—than I was the day I finished reading *Microbe Hunters* 54 years ago. I believe we are on the threshold of a revolution in medical microbiology that will be the most important achievement in human medicine since the discovery of the germ theory of infection by my beloved microbe hunters of the past. We will, however, have to learn to deal with data in a new way and to exchange information in ways that current journal formats and other means of communicating science cannot accommodate. In a broader vein, microbiologists have at their fingertips the genetic information for a staggering number of the most abundant life forms on the planet and the most biological diversity to explore. The exploration of the biology of microbes will be equivalent to thousands of human genome projects. There is a line from Dr. Seuss that seems appropriate here. It goes something like "Oh the things we shall see. Oh the places we shall go."

STANLEY FALKOW was born in Albany, New York, in 1934. He received his bachelor of science degree from the University of Maine and his master of science and doctoral degrees from Brown University. He has taught and done research at several universities and hospitals and served as Chairman of the Department of Medical Microbiology at Stanford University. He was the President of the American Society for Microbiology from 1996 to 1997. He is currently Professor of Microbiology at the School of Medicine at Stanford University. He is a member of the American Academy of Arts and Sciences, the National Academy of Sciences, and the Institute of Medicine. He has also been the recipient of the Paul Ehrlich-Ludwig Darmstaedter Prize, the Becton Dickinson Award in Clinical Microbiology, and the Altmeier Medal from the Surgical Infectious Diseases Society of America.

The following papers are representative of his publications:

Monack, D. M., J. Mecsas, D. Bouley, and S. Falkow. 1998. *Yersinia*-induced apoptosis *in vivo* aids in the establishment of a systemic infection of mice. *J. Exp. Med.* **188:**2127–2137.

Cirillo, D. M., R. H. Valdivia, D. Monack, and S. Falkow. 1998. Macrophage-dependent induction of the *Salmonella* pathogenicity island 2 type III secretion system and its role in intracellular survival. *Mol. Microbiol.* **30:**175–188.

Finlay, B. B., and S. Falkow. 1997. Common themes in microbial pathogenicity revisited. *Microbiol. Mol. Biol. Rev.* **61:**136–169.

Valdivia, R. H., and S. Falkow. 1997. Fluorescence-based isolation of bacterial genes expressed within host cells. *Science* **277:**2007–2011.

Bliska, J. B., J. E. Galan, and S. Falkow. 1993. Signal transduction in the mammalian cell during bacterial attachment and entry. *Cell* **73:**903–920.

Relman, D. A., T. M. Schmidt, R. R. MacDermott, and S. Falkow. 1992. Identification of the uncultured bacillus of Whipple's disease. *N. Engl. J. Med.* **327:**293–301.

Bliska, J. B., K. L. Guan, J. E. Dixon, and S. Falkow. 1991. Tyrosine phosphate hydrolysis of host proteins by an essential *Yersinia* virulence determinant. *Proc. Natl. Acad. Sci. USA* **88:**1187–1191.

Relman, D. A., J. S. Loutit, T. M. Schmidt, S. Falkow, and L. S. Tompkins. 1990. The agent of bacillary angiomatosis. An approach to the identification of uncultured pathogens. *N. Engl. J. Med.* **323:**1573–1580.

Lee, C. A., and S. Falkow. 1990. The ability of *Salmonella* to enter mammalian cells is affected by bacterial growth state. *Proc. Natl. Acad. Sci. USA* **87:**4304–4308.

Arico, B., J. F. Miller, C. Roy, S. Stibitz, D. Monack, S. Falkow, R. Gross, and R. Rappuoli. 1989. Sequences required for expression of *Bordetella pertussis* virulence factors share homology with prokaryotic signal transduction proteins. *Proc. Natl. Acad. Sci. USA* **86:**6671–6675.

Miller, J. F., J. J. Mekalanos, and S. Falkow. 1989. Coordinate regulation and sensory transduction in the control of bacterial virulence. *Science* **243:**916–922.

2

CYNTHIA A. NEEDHAM

Wednesdays

Wednesdays always begin with Bert firing up the espresso machine. This is no easy task. The machine is irascible, and Bert is now so short of outlets in his studio that the simplest need for extra electricity requires reorganizing the complex bank of plugs supporting our various computers. The weekly latte production is only one of many concessions that Bert has made to The Book. The onslaught of computers is one of the others.

Mahlon paces before the boards, now festooned with the pages of words and Bert's drawings that will become The Book. When I watch him, I always think he seems surprisingly comfortable with this process for a scientist, and my respect for him deepens. I also sometimes think that at this stage of his life he should be at the helm of a cherished sailboat, face to the wind, rather than pacing before a room-sized set of boards in a stuffy studio in Vermont.

The group is complete when Kenneth joins us. He has chosen to make his life with me as both husband and intellectual partner. Sometimes this is a painful combination. He has forced the three of us to confront every problem as if it were a mathematical proof, imposing rigor where we had created beauty but fell short of logic. I think The Book and I are better for his persistence, but it is not always easy. It is only on reflection, outside the heat of debate, that I have arrived at this conclusion.

Kenneth and I are new at this; Bert and Mahlon are veterans. Their first struggle over truth and beauty yielded a most wonderful book: *The Way Life Works*, a work I recommend to all. Kenneth and I are grateful that they have joined us, for these are daunting days, and their experience has helped guide us all through some of the perils of this process.

Bert is an acclaimed artist and illustrator. Mahlon is a distinguished senior scientist. Kenneth is a slightly undisciplined mathematical genius. You may wonder why such a seemingly disparate group has ended up together, sharing every Wednesday regardless of weather, family, and personal commitments. It is because we are all motivated by the same vision—a general public with a high level of scientific literacy. This means that we all share the same commitment—one of translating scientific discovery and principles into terms and stories that are intellectually accessible to everyone. And that is what The Book is about.

The Book and what it stands for have occupied a good part of my life for over a year. We have collectively examined every word, negotiating between the scientific "truth," as we perceive it, and the desire to communicate the beauty and power of the unseen world of microbes to people who don't know they exist. For that is also what The Book is about, and that is why we are always together on Wednesdays.

My Wednesdays with Bert and Mahlon and Kenneth have come at the beginning of a new path, a path for which I left a successful career as a scientist and administrator in a large New England medical system. The Book is only a part of my departure. I have had the privilege of leading a coalition of organizations known as the Microbial Literacy Collaborative on behalf of the American Society for Microbiology. The goal of the Collaborative was to increase the level of scientific literacy among the members of the general public, with a particular focus on the microbial world. The centerpiece of the initiative's effort is a four-hour documentary broadcast by PBS entitled *Intimate Strangers: Unseen Life on Earth*. The members of the Collaborative also created a twelve-part video series for classroom use, *Unseen Life on Earth: An Introduction to Microbiology*, a set of inquiry-based activities for use in both formal and informal educational settings, and Microbeworld.org, an online resource for students, parents, and teachers. Members of the Collaborative conducted youth summits to introduce the microbial world and the available resources from the initiative to youth leaders from across the United States. And of course, there is The Book.

My commitment to building scientific literacy is one shared by many scientists, but few have the opportunity that I have been given. The success of the Microbial Literacy Collaborative's initiative is due in large part to the combined energy of talented producers, educators, and scientists. All brought their strengths and perspectives together to articulate an area of science that touches the lives of all and about which there exists much public ignorance and apprehension. Without the unwavering commitment of the scientists, however, this effort would have been far less successful. Their insights and advice provided the key link between the leading edges of research and its application and the translation of this highly technical area into accessible language and imagery.

We are about to enter an age when scientific illiteracy will no longer be an option for the citizens of the world. Almost every decision—whether it is private or public, whether it affects only one individual or many individuals within a community, state, nation, or our global village—will depend on some knowledge of science. Nonscientists and scientists alike will face a daunting array of questions. Should we take antibiotics if we are uncertain of their effectiveness? Should we vote for a representative who has written legislation to block the use of genetically engineered crops? Should we support or oppose the release of novel microorganisms to clean a toxic spill? What is the best way to prevent loss of the ozone layer? Is global warming a reality or a myth, and are we willing to pay more at the gasoline pump to recognize the social costs of our transportation choices?

Popular media such as television, radio, print, and the Internet provide the opportunity to build a more comprehensive appreciation for both the scientific process and the contemporary scientific basis for understanding complex issues. The success of these media in expanding science literacy will depend, however, on input from the scientific community. It is critical that we shed our traditional distrust of the very mechanisms that reach the public most effectively and participate in ensuring the quality and scientific integrity of the messages they convey.

That is why Wednesdays are so special. Wednesdays are how I know I have chosen the right path. I am elated about my journey and encourage all who care about the well-being of our citizenry to join me.

CYNTHIA NEEDHAM received her doctoral degree in microbiology from Oklahoma State University in 1974. She immediately accepted a position as postdoctoral fellow in the Department of Microbiology and Infectious Diseases at Creighton University School of Medicine in Omaha, Nebraska, and a year later became Assistant Professor in that department. In 1976 she became Director of the Clinical Microbiology Laboratory at St. Joseph's, the principal teaching hospital for Creighton University Medical Center. In 1980 she moved to Boston to accept a position as Director of the Microbiology Laboratory of Boston University Medical Center and to join the faculty of the Department of Microbiology at Boston University School of Medicine. Five years later she was recruited by Lahey Medical Center to become Head of Microbiology; there she ultimately assumed responsibility for business development and management of all external laboratory services. In 1987 she joined the adjunct faculty at Tufts University School of Medicine. She also is an associate professor at Boston University School of Medicine. She was elected National Secretary of the American Society for Microbiology (ASM) and served from 1988 to 1994. She helped formulate ASM's strategic plan for public communication and assumed the Chair of the Executive Board of the Microbial Literacy Collaborative. In this role she helped obtain multimillion dollar funding for the microbial literacy effort. In 1998 she resigned from Lahey and founded ICAN, a company that develops broad-based public education initiatives in science, technology, and business, with a particular interest in areas where science and public policy intersect. Under her leadership ICAN has partnered with Oregon Public Broadcasting to bring science to the public. She is President of ICAN (and secretary, cook, and chief bottle washer). ICAN is not an acronym, but it does have a double entendre—I Cynthia A. Needham and I CAN.

The following papers are representative of her publications:

Needham, C. A., K. A. McPherson, and K. H. Webb. 1998. Streptococcal pharyngitis: Impact of a high-sensitivity antigen test on physician outcome. *J. Clin. Microbiol.* **36:**3468–3473.

Needham, C. A., and P. Hurlbert. 1992. Evaluation of an enzyme-linked immunoassay employing a covalently bound capture antibody for direct detection of herpes simplex virus. *J. Clin. Microbiol.* **30:**531–532.

Needham, C. A. 1988. *Haemophilus influenzae*: Antibiotic susceptibility. *Clin. Microbiol. Rev.* **1:**218–227.

Serfass, D. A., P. M. Mendelman, D. O. Chaffin, and C. A. Needham. 1986. Ampicillin resistance and penicillin-binding proteins of *Haemophilus influenzae*. *J. Gen. Microbiol.* **132:**2855–2861.

Needham, C. A. 1986. Rapid methods in microbiology for in-office testing. *Clin. Lab. Med.* **6:**291–304.

Sweeney, K. G., A. Verghese, and C. A. Needham. 1985. *In vitro* susceptibilities of isolates from patients with *Branhamella catarrhalis* pneumonia compared with those of colonizing strains. *Antimicrob. Agents Chemother.* **27:**499–502.

Needham, C. A., and W. Stempsey. 1984. Incidence, adherence, and antibiotic resistance of coagulase-negative *Staphylococcus* species causing human disease. *Diagnost. Microbiol. Infect. Dis.* **2:**293–299.

Section Two

VISION ON A QUEST

"When you look back at them, many of the fundamental discoveries of science seem so simple, too absurdly simple. How was it men groped and fumbled for so many thousands of years without seeing things that lay right under their noses? So with microbes…" So writes Paul de Kruif in *Microbe Hunters*.

When Antony van Leeuwenhoek was born in 1632, there were no microscopes, no tools to pry open the microbial world that surrounds us. Van Leeuwenhoek was not a scientist; he was a cloth maker, tailor, and owner of a dry goods store who used magnifying lenses to view fabrics. He was also a surveyor and official wine taster of Delft, Holland. At that time, it was not fashionable to pursue careers in science. Van Leeuwenhoek followed the traditional path to his trades and to his life in the community. Philosophers and theologians held center stage, and mysticism and the mysteries of life prevailed. Although deeply religious, van Leeuwenhoek had a love affair with exploring—with searching for an unseen world. Thus, by flickering candlelight, he searched for images and things that had never been seen. He was not deterred by the tedious; in fact, he was habitually drawn to hours of grinding lenses and seemingly interminable peering through his primitive, dimly lit microscopes. But to see what? "Little animalcules" moving about, or only the figments of his imagination?

To pursue the invisible world, van Leeuwenhoek must have been a maniacal observer. Who else would have turned his lens on clear, pure rainwater? And who else but such a religious and fastidious man would have tried to find out whether God sent little animals to earth in the rain? Was water only water—or was it a soup of microbes? So it was that van Leeuwenhoek recorded the first drawings of the microbial world. But other, more learned men would have to recognize the significance of what he saw, to make sense of what he communicated in his many long, rambling letters to the Royal Society of London. And that was not easy, for the world of microbes was so strange, its discovery so surreal. Even Carl Linnaeus, the Swedish botanist who developed a universal system for classifying plants and animals, said of van Leeuwenhoek's microbes: "[T]hey are too small, too confused, no one will ever know anything exact about them, we will simply put them in the class of Chaos."

So for centuries, most microbes and the meaning of the real microbial world escaped us. How things have changed, how deeply we can look with electron beams and gene probes, how great our knowledge grows. And with the Internet, fax, and cellular phone, how fast we can tell the world of the changes we see.

Discovering the Real Microbial World

Not surprisingly, no one—not even a scientist such as myself—starts out interested in microorganisms. Our first encounters with microorganisms are usually unpleasant. When you're a kid, some witch doctor of a pediatrician sticks a needle in you and mumbles something about "Now you won't get...." Get a sore throat and a high fever and it's the doctor again, telling you and your mom about a streptococcus making you feel bad and you must take your antibiotic to get rid of the bad little bugs. If your milk tastes bad, someone's likely to tell you little "organisms" you can't see are making it that way. Ask your father why one of the bottles he's keeping in the cellar exploded, and he starts in again about little "organisms."

Finally, when you're well into school, some biology teacher will show up with some small, flat, round, covered dishes called "Petri plates" containing a jelly-like substance and open one to the air, put a drop of water on another, and let you put the most yucky thing you can think of on a third. After a day or two, all those plates (except the ones the teacher didn't open) have stuff growing on them, which you are told are bacterial colonies. No wonder you couldn't see them if it takes more than ten thousand to make up one little spot on the Petri dish. And then the big day comes when the teacher introduces you to a microscope and mixes a little bit of one of those colonies with a drop of water on a slide, focuses the thing, and lets you look. There they are! Thousands of them. Now all that stuff you've heard when you were young finally has something tangible behind it. This is the first time you get to sense microorganisms in their own right, not just in terms of what they do to you or for you. Just imagine how the Dutchman Antony van Leeuwenhoek must have felt, just over 300 years ago, when he made the first primitive microscope and was the first person ever to see microorganisms, a whole world of living things no one knew existed! Van Leeuwenhoek found them everywhere, in everything, all different sizes and shapes, some colored, some not, darting around, others just sitting there and jiggling (Brownian motion). Imagine all the thoughts and questions that must have gone through van Leeuwenhoek's mind!

Yet despite all the variety and wonder in the microbial world, microbiology is the poor relative of the biological sciences, the "ho-hum" branch of biology. With a few (very important) exceptions, the scientists it tends to attract are usually pack rats—biologists who love to collect little stories about everything (that signify nothing), not caring how it all fits together into a big picture. You also get types I call "technological adventurists" who demonstrate their scientific prowess by finding new and powerful ways to destroy disease-causing bacteria (which has some merit). Then there are scientists who find bacteria to be convenient systems in which to study "basic biological problems." In none of these cases is there a genuine interest in microorganisms or in understanding bacteria and the microbial world. Microorganisms provide quaint little anecdotes; they have to be controlled; they are useful systems for other studies. That's all.

Microbiology today attracts relatively few students. Most seem to prefer the glamour, excitement, and passion of working with animals and plants or choose to skip organisms altogether and cruise the high-tech boulevards of the molecular world. University administrators nowadays like to close down microbiology departments: One or two lectures on bacteria in a general biology class is enough; the students get bored with a lot of trivial junk about where this bug was isolated and what it grows on; the understanding's all on the molecular level anyway! If they want organisms, give them something they can relate to, like spotted owls, whales, leopards, the rain forest, and so on—something that's "relevant"!

Why? Why is this? Why has there been so little interest in microorganisms by scientists and the general public? There is a good reason, but it's not what microbiologists think it is. It has nothing to do with microorganisms being small so we can't readily see and definitely can't relate to them. Look at the interest that certain viruses spark, not only among scientists but also in the general public. Look at how many people are interested in molecules. Both are far smaller, and less cuddly, than bacteria. And the reason couldn't be because microorganisms are relatively unimportant or insignificant in the "big picture."

Microorganisms *are* the big picture! They not only constitute most of the biomass on this planet, they also sit at the base of most of the food chains and are ultimately responsible for most of the recycling of biomass. They are directly or indirectly responsible for the creation and maintenance of our oxygen-containing atmosphere. Remember, a plant is a plant only because it contains chloroplasts, and chloroplasts have bacterial ancestors. Microorganisms are central to the carbon balance and even are dynamic forces in mineral deposition.

So what is the reason for the general lack of interest in microorganisms per se? Why is microbiology's lowly status among the biological sciences so out of keeping with the dominance of microorganisms in the biosphere and, thus, the importance of studying them? The problem lies with us—we professional microbiologists, whose job is to understand microorganisms and to communicate that understanding, its importance, and its beauty to students and the public in general. But it's not for lack of talent or desire that we have failed you. If we don't understand microorganisms, which we didn't until very recently, how could we help anyone else understand and appreciate them? Some of you, and probably a lot of microbiologists, are asking, "How can that guy say that microbiologists did not understand microorganisms?" The answer turns on what it means to understand something in biology.

Knowing a lot of facts about an organism is not enough. The essence of understanding an organism lies in understanding how it relates to other organisms. Life is an interconnected web of organisms, and the main struts—the main fibers—in this web are evolutionary relationships. Understanding an organism in a biological sense means understanding it *and* its place in the web of life. That is why comparative anatomy, for example, is so important. You can't begin to study ecology without such a web. We can stroll through the woods and delight in distinguishing among various animals, birds, plants, and mushrooms. The web is the framework within which this is all possible, all meaningful. Imagine how different our stroll would be if we couldn't comprehend all birds as of the same kind, if we didn't know whether an elephant might not be as related to a hummingbird as is a condor. This example is absurd, but only absurd in the world of animals and plants. It typifies the situation in microbiology! This is what caused two great microbiologists, C. B. van Niel and Roger Stanier, to characterize as an intellectual scandal our failure to have a decent concept of bacteria. You see, microbiologists couldn't tell in general whether two bacterial isolates were as closely related as rats and mice are or as distantly related as birds and elephants. But it was worse—far worse, it turns out. You couldn't even tell whether two bacterial strains were more distantly related than animals are to plants! Except in trivial cases, microbiologists had no idea whatsoever of the natural relationships among bacteria. We just couldn't generate any of the main fibers in that critical web of connectedness, of understanding, for the microbial world.

Today we finally can! For the first time, microbiologists are developing a biologically meaningful understanding of microorganisms. Now we can begin to tell everyone how beautiful, fascinating, really important, and diverse the microbial world is.

My own work has been central to this critical change in microbiology, for I had the good fortune to be the right person in the right place at the right time when it all started. Initially, I had no interest in microbial relationships. Indeed, in one sense I wasn't even aware of microbiology's plight, of the heroic but vain attempts to determine microbial phylogenies and of the counterproductive way microbiologists had come to rationalize their failure to do so. Microbiologists of the third quarter of the twentieth century seemed to have concluded that it was impossible—innately impossible—to determine a comprehensive bacterial phylogeny, but that it was all right because such a phylogeny wasn't necessary to understand bacteria. All we needed to do was figure out how bacteria (prokaryotes) differed from eukaryotes. Imagine what a concept of animals we would have if we did not know their relationships to one another but only knew how they as a group differed from plants! Anyway, at the time (late 1960s) I was a biophysicist turned molecular evolutionist, and my scientific passion was to understand the genetic code, which I'd come to realize was part and parcel of a larger question—that of how the translation apparatus evolved. But you couldn't begin to answer questions about the evolution of the genetic code then because a universal phylogeny showing the evolutionary relationships of organisms is the essential framework within which to deal with any such question. Obviously no such framework existed—not by a long shot.

In hindsight, my perspective on this issue was a unique one. Many molecular evolutionists of that time knew that the technical power existed in molecular sequencing to determine distant phylogenetic relationships. Some understood that with the right molecules—for example, transfer RNAs—the full breadth of phylogeny could be spanned. Yet from their actions, it appeared that none of them appreciated the importance of determining a universal phylogenetic tree, what powerful evolutionary knowledge would emerge therefrom. On the other hand, microbiologists, who in the past had understood the importance of knowing phylogenies, at least in regard to classifying and to understanding bacteria, now believed that determining a comprehensive bacterial phylogeny—not to mention a universal one—was impossible. I alone, it seems, felt the importance of having a universal phylogeny, and I knew how it could be determined—by comparing the sequences of a functionally universal and highly conserved molecule such as ribosomal RNA (rRNA). Without that phylogeny, we didn't have a prayer of getting at the deep evolutionary questions about the origin of the cell. The rest of the story is pretty much a matter of record.

The program of determining a universal phylogeny through rRNA sequences got off to a slow but promising start. In the early 1970s, it wasn't feasible to sequence a molecule as large as 16S rRNA, so we had to settle for sequencing little pieces of it. The big break came when Norm Pace's lab worked out a method for sequencing over 95% of a 16S rRNA molecule in a day or two. Then we really were able to move forward. Many spectacular findings came out of our program, as was bound to be the case when a comprehensive microbial phylogeny could be finally determined.

Almost all the old criteria used to relate bacteria were shown to be the disasters earlier microbiologists had feared they were: Morphology proved an almost worthless phylogenetic measure, and physiology (with a few exceptions) was only a little better. Myxobacteria turned out not to be separate from the other bacteria at the highest level, as had been thought. And *Beggiatoa* turned out not to be *Oscillatoria* that had lost photosynthetic capacity, as almost all microbiologists had come to believe. Yet the mycoplasmas, which most microbiologists took to be very distinct phylogenetically, were no more than a subgroup of degenerate clostridia. The old "gram-negative" taxon fell apart completely. Gram-positive turned out to be a somewhat telling phylogenetic measure, but it misses many members of the real taxon, including the all-important photosynthetic "gram-positive" bacteria. And on and on.

The most spectacular single finding to come from our program—the one that would astound us all—was the discovery of the archaea, bacteria that are not bacteria and that have no relationship to ordinary bacteria. Microbiologists and other biologists had come to see all bacteria as a kind, all coming from a common ancestor distinct from the common ancestor that gave rise to the eukaryotes. So firm was the belief in this dogma that the vast majority of microbiologists could not bring themselves even to contemplate that there might be two completely separate groups of prokaryotes, that the Archaebacteria (as they were then known) were unrelated to normal bacteria. Thus you can imagine the scoffing that went on when we made such an outlandish claim in print, and the feathers that got ruffled when we persisted in our claim in a not-so-gentle fashion. It wasn't that we took any great pleasure in an "in-your-face" approach to presenting our case, but when wrong ideas are deeply ingrained and have become institutionalized, as was the case here, you just don't change them by coming with hat in hand and saying, "Gee, look what we found."

So, what are these archaea? Why should anyone take special interest in them? What has their discovery done to change our view of the microbial world? And how are they changing the science of microbiology and biologists' view of it (we hope)? The background against which questions of this sort have to be viewed and answered is the old microbiology—microbiology as it was before the discovery of the archaea, the microbiology that was the bottom rung of the pecking order of biological sciences.

The old prokaryote-eukaryote dichotomy was short on understanding and long on dividing biology into two camps, which more or less went their separate ways. The emphasis was on the ways in which prokaryotes and eukaryotes differ, not on why they differ and what sort of common ancestor they share. In other words, the emphasis was on pigeonholing and distinguishing, not on synthesizing and understanding.

That the archaea are prokaryotic is an almost meaningless statement. It basically means they don't appear to have nuclear membranes and organelles. So what? What does that tell us about them? To understand what it is to be bacterial, archaeal, or eukaryal, you have to operate almost exclusively on the level of molecules. And when you do this, you don't see prokaryotes on one side, eukaryotes on the other. You see three different types of organisms, related in ways but each highly unique. It is not that the archaea, or either of the other primary groupings of organisms, do things in an entirely novel way. It is in the exact way they perform the universal functions of a cell and in the details of the design of the cell itself that the three domains distinguish themselves from one another. Although all three came from a common ancestor, that ancestor (I won't go into its possible nature here) was certainly rudimentary by the standards of modern living systems. At this ancestral stage, life was cellular, but cells were primitive; they had not yet become the fully evolved modern cells with which we are familiar. That is why the archaea, bacteria, and eukaryotes are so different; each represents an independent solution to the problem of how to make (evolve) a "modern" cell. The discovery of the archaea proved especially important for forcing us to focus on the central question of how cells evolved in the first place. How did the three primary lines of descent come to be?

Speaking crudely, cells have a functional aspect and a replicative aspect. In the functional aspect, cells are metabolic machines with multitudes of different proteins, catalyzing this and that, building various structures, transporting things from here to there. The replicative side is centered about the chromosome, the queen bee of the cell, responsible ultimately for the continuation of life and tended by a court of special molecules: polymerases, gyrases, repressors, histones, and repair system enzymes. In their metabolic machinery, the archaea and bacteria resemble one another. They seem biochemical giants compared to the eukaryotes, and their biochemistries at first glance appear to have much in common. The archaea are unique in such things as their capacity to make methane (the methanogens); have ether-linked, branched-chain lipids rather than ester-linked, straight-chain ones; and have a unique and very simple photochemistry (seen in the extreme halophiles) based on bacterial rhodopsin.

Yet the archaea are remarkable for their capacity to metabolize at high temperatures. Some grow optimally even above the normal boiling temperature of water, and some of their enzymes in isolation can be active at temperatures still higher than this. The bacteria, on the other hand, deserve the molecular Nobel Prize for having invented the energy source on which life as we know it basically rests: chlorophyll-based photosynthesis. Thus, from the metabolic perspective, archaea and bacteria resemble one another. A microbiological biochemist might say, "So, what's the big deal?"

Go look at the chromosome and its molecular entourage, however, and a different picture emerges. There are major differences in the replicative functions of archaea, bacteria, and eukaryotes. Although the cell's translation mechanism is highly conserved universally, each of the three primary groups of organisms has its peculiar variation of it. And here we see a different picture of relationships emerging. Among the ribosomal proteins, there exist subsets that are found only in bacteria or eukaryotes. The archaea seem not to have homologs of the former, but a number of homologs of the latter. Similarly, in their aminoacyl-tRNA synthetases, the archaeal versions tend to be more similar to the eukaryotic versions than to the bacterial ones. Turning to transcription, the case is even clearer. The bacterial RNA polymerase is relatively simple, comprising three different subunits, with a fourth involved merely in initiation of the process. Eukaryotes and the archaea, however, have a more complex transcription mechanism involving the basic three subunits and several smaller ones, many of which are homologous between the two groups. Two different modes of transcription initiation exist: one in the bacteria and another in the archaea and eukaryotes. Then, with the "ultimate" process—the replication of the genome itself—the story repeats. The archaeal version is far more like that found in eukaryotes than that seen in the bacteria. What this suggests is that the basic rudimentary machinery for cell replication that existed in the universal ancestor was refined separately on the bacterial line of descent and on an ancestral line of descent shared by the archaea and eukaryotes. Further independent refinements, of course, occurred on both the individual archaeal and eukaryotic lineages. This is the strongest evidence to date that, in some respects, at least the archaea and eukaryotes share a common evolutionary heritage and that the archaea can be viewed as direct descendants of what biologists believe to be the prokaryotic ancestor of the eukaryotes.

The archaea are central to a real revolution that is still occurring in microbiology. They epitomize the importance of phylogeny and the power of molecular biology to reveal evolutionary history. From a knowledge of microbial phylogeny flows (1) a microbiology that finally has a useful concept of bacteria; (2) a microbial ecology that can now define the chemical dynamics of a microbial niche in organismal terms; (3) a structure within which the voluminous collection of anecdotal facts about bacteria can be given some useful meaning; (4) a solution to microbiology's great chronic problem—the inability to know anything about uncultured organisms (the vast majority of microbial species are not, and possibly cannot be, cultured); (5) a new and far more productive view of the relationship between prokaryotes and eukaryotes; and (6) a microbiology that has become an evolutionary discipline—indeed the forefront of future evolutionary research.

From all this will ultimately come a time when microbial biology assumes its rightful place in the pantheon of biological sciences as the leading discipline—a position commensurate with the place microorganisms hold in the natural order of things—a place I will have helped find.

CARL WOESE was born on July 15, 1928, in Syracuse, New York. He received his doctoral degree from Yale University in biophysics. He is a member of the National Academy of Science. He currently is Stanley O. Ikenberry Professor of Microbiology at the University of Illinois. Important developments in his career include the discoveries (circa 1965) that the genetic code is central to understanding the nature of life; that understanding the genetic code is one of the great evolutionary questions; and that reconstructing the universal phylogenetic tree is the essential framework within which to answer it. These discoveries set the course of his career, from which he has never deviated. The milestones of that course are (1) defining the amino acid polar requirement (1966); (2) creating a molecular model for translation (1970); (3) discovering the basic phylogenetic structure of all life (1977, 1987); (4) elucidating the secondary structure of rRNA (1981); and (5) uncovering the nature of the universal ancestor (1982, 1998). Mitchell Sogin studied with Woese and is one of his protégés carrying forth the quest to extend understanding of phylogenic relationships of microorganisms.

The following papers are representative of his publications:

Woese, C. R. 1998. The universal ancestor. *Proc. Natl. Acad. Sci. USA* **95:**6854–6859.

Woese, C. R. 1987. Bacterial evolution. *Microbiol. Rev.* **51:**221–271.

Noller, H. F., and C. R. Woese. 1981. Secondary structure of 16S ribosomal RNA. *Science* **212:**403–411.

Fox, G. E., E. Stackebrandt, R. B. Hespell, J. Gibson, J. Maniloff, T. A. Dyer, R. S. Wolfe, W. E. Balch, R. S. Tanner, L. J. Magrum, L. B. Zablen, R. Blakemore, R. Gupta, L. Bonen, B. J. Lewis, D. A. Stahl, K. R. Luehrsen, K. N. Chen, and C. R. Woese. 1980. The phylogeny of prokaryotes. *Science* **209:**457–463.

Woese, C. R., and G. E. Fox. 1977. Phylogenetic structure of the prokaryotic domain: The primary kingdoms. *Proc. Natl. Acad. Sci. USA* **74:**5088–5090.

Woese, C. R. 1970. Molecular mechanics of translation: A reciprocating ratchet mechanism. *Nature* **226:**817–820.

Woese, C. R., D. H. Dugre, W. C. Saxinger, and S. A. Dugre. 1966. The molecular basis for the genetic code. *Proc. Natl. Acad. Sci. USA* **55:**966–974.

BRENDLYN D. FAISON

Searching for the Meaning of "Life"

A child of a steelworker and a first-grade teacher. A first-generation Yankee from northern Ohio raised in a steel town. Born to parents who had fled the South—lured to the Rust Belt by the promise of good jobs and racial equality. Freed to pursue knowledge by *Brown v. Board of Education*. Given just opportunity by U.S. Supreme Court rulings. Benefactor of the Presidential "Great Society" programs. And fortunate not to have recognized just how difficult it is to define "life."

I see myself, a girl of 6 standing in front of my mother's refrigerator, asking if each item before me was alive or dead. What was it to be alive? The grapefruit was alive, but I didn't appreciate that—at least not then. A child looking for answers. What was life? Angered by circular dictionary definitions—what did it mean to say, "Life is a quality possessed by living things"? What was it to be really alive? Blind faith seemed too easy. Television coverage of the Mercury, Gemini, and Apollo astronauts going forth in search of new life forms only fueled my questions. What were they looking for? (Or was it Captain Kirk, Mr. Spock, and Doctor McCoy on *Star Trek*?) Whoever it was, it only caused me to question more the definition of life—to struggle to figure out how to recognize that which was, is, or could become alive.

Now 16 years old and in need of education to find my way. Time to let go of my days as a tomboy. Prodded by my high school guidance counselor to attend a liberal arts "Seven Sister" college in the hope of giving me "some polish" (his words). Trusting in fate, I flipped a coin and went to Wellesley. However, I had been subtly encouraged throughout public school to avoid science and mathematics as fields "not becoming to a lady." (Let's not even address the racial issue.) Time to put aside my questions of the meaning of life. Time to find my own life. Time to major in French literature.

But Wellesley required that all students take at least three science courses of the sixteen required for the degree in any field; even French majors need to appreciate science. One of those science courses had to have a laboratory. So I registered for Biology 104—microbiology, a choice based on expediency because it fit the hole left in my schedule between my language and philosophy courses. I rationalized that I would be forgiven the "D" I expected to earn. It was my first time away from home and I was so young. Little did I know that the lab work would be fun and that the course instructor, Mary M. Allen, would become my role model.

I never got the hang of mitosis and meiosis—or of eukaryotes in general—so subsequent macroorganismal and cell biology courses did not go well. Fortunately, Wellesley had an exchange program with MIT. I took the remainder of the courses for what was now my major (biological sciences, emphasizing microbiology) and met my mentor, Arnold L. Demain, at MIT. Somewhere along the line I realized it would be easier for me to major in biology and take "minor" courses in chemistry, physics, and mathematics than to major in French and take ancillary courses in history, religion, and psychology. So I did.

And I came to realize that biology was but a subset of chemistry, physics, and mathematics, with rules and regulations that could be flouted or exploited according to circumstances. I constructed a three-dimensional binary matrix of bacterial genera capable of chemotrophy versus phototrophy, lithotrophy versus organotrophy, and autotrophy versus heterotrophy. Most emphasis in microbiology at that time was on human pathogens—metabolic chemoorganoheterotrophs that were the links in a food chain anchored ultimately in photosynthesis. Somehow my 2^3 summation suggested that there were lots of different metabolic lifestyles. Living to see the current plethora of new genera is pleasing. My thoughts were right. Those microbes living in hostile extreme ecosystems, those living in the near-starvation conditions of oligotrophic lakes, those living in the deep subterranean dark biosphere all have distinct metabolisms.

Now I see myself a college graduate. Bowing to familial and societal pressure, I go to work as a lab technician at Albert Einstein College of Medicine, pretending that I would use my job as an entrée to medical school. But a career in medicine has *never* interested me. I still hadn't figured out life. Well, in intact humans, I pretty much had; individual cells were a different issue. Perhaps my problem was that at Einstein I worked first on plant and animal iron-based hemoglobins—a good introduction to inorganic biochemistry but not necessarily to organismal life—and then I worked on DNA-based adenovirus 12 and related strains, none of which is quite alive. To understand life on a true cellular level, I needed more education. I was anxious to move on.

So I enroll in the graduate program at MIT. I do so primarily because of Arnold Demain, who was not fated to be my thesis adviser, and because Boston and Cambridge were familiar. I never noticed that MIT was an engineering school. I am not an engineer—I never wanted to be one. So I entered the School of Science. But the School of Science focuses on DNA. For reasons I still cannot understand, I have little if any interest in nucleic acids. I like proteins instead and was later delighted by the discovery of prions. I was exhilarated by the opportunity at MIT to work with an extremophile (*Clostridium thermocellum*).

Times are changing and so is Cambridge. The DNA-based biotechnology revolution is growing, and Cambridge is at the heart. Random violence accompanying the forced desegregation of the Boston-area public schools is close by. Cambridge is a little uncomfortable for me. Moreover, I am in danger of being forcibly molded into a chemical engineer, and a barely competent one at that. I am impelled to pursue my true love, organismal microbiology. So on a bleak January morning (20° below zero, with a wind chill of 80° below zero), I arrive at the University of Wisconsin in Madison—*sans* adviser, *sans* apartment, *sans* decent winter coat.

Thanks to Einstein, who I am sorry to admit was not my patron, I had noticed that energy, matter, and mathematics seem to define existence. My BS-level microbiology studies had taught me that respiration, or energy metabolism, may be defined as the reduction of specific chemical elements (O>N>S>C, with respect to energy yield). T. Kent Kirk, my Ph.D. thesis adviser (whose research team was housed at the U.S. Department of Agriculture's Forest Products Laboratory on the UW campus), taught me that oxygen can be reduced incompletely to form extremely energetic degradative catalysts. Indeed, my thesis described the regulation of active oxygen production by the fungus *Phanerochaete chrysosporium*. Nonetheless, my degree is in bacteriology, and I am pleased to have had Jerald S. Ensign as my on-campus patron saint. My only regret is that I never accompanied him on his annual "iron bacteria hunt," during which his students learned the potentially key role of iron bacteria in biochemical transformations in the environment.

I now have a Ph.D. Time to commence my career in microbiology. So I take a job at Procter & Gamble to work for Robert J. Larson, a University of Wisconsin graduate. My job was to test the biodegradability by mixed bacterial cultures of consumer product ingredients such as those used in fabric softeners, laundry detergents, and diapers. The real world of microbiology was very different from the one I had expected to inhabit. I had begun my microbiological career by studying the degradation of glucose as in typical class exercises involving heterotrophs. My first undergraduate research project had explored use of carbon dioxide and/or acetate by Dr. Allen's cyanobacteria. At MIT, I had worked on cellulose degradation; at the University of Wisconsin, I stepped into the world of ligninolysis. All of the materials I had studied in school were natural. Now at Procter & Gamble, I tried to prove that man-made products destined to end in landfills, septic systems, and sewage treatment plants were indeed biodegradable. It was an interesting challenge.

But after only two years of consumer products research—and repeated threats of transfer into management—I was persuaded to leave private industry to do bioengineering (!) research at Oak Ridge National Laboratory (ORNL). My position at ORNL was as a microbiologist in support of a coal conversion process development effort led by Charles D. Scott. Hey, there is a lot of coal in the world; it is biodegradable, but only in the presence of oxygen. The fungus *Paecilomyces* works well, although the higher basidiomycetes are preferable. My studies of the oxygenative breakdown of lignin by fungi had proved relevant.

The Department of Energy (which owned ORNL) also faces problems with radioactive environmental contaminants. So I developed an independent program on microbial remediation of anoxic environments contaminated with uranium. I like bioprocess development, particularly when I can help discover and exploit novel biological energy conversions. To my surprise, the process I studied revealed the respiration of uranium, bound to oxygen, by *Pseudomonas aeruginosa*—and led me back to the concept of iron as a key mediator of electron transport (and oxygen manipulation) in biological systems.

Things are just getting interesting in the laboratory when I have a dreadful automobile accident. Lying in a hospital bed, I have lots of time to contemplate all I have learned about "life." It is a time to think about life and for my thoughts to ramble. My training as a scientist has told me that life is simply a display of electromagnetism—the energy form critical to life and inherent to matter—and can be reduced to simple assessments of electron flow and stability. Mathematics, specifically probability, indicates that carbon is a good basis for life *as we know it* because of carbon's abundance in the universe. The concentration of all elements decreases exponentially with atomic number, except for iron. Iron is the ultimate decay product of very-high-

THE BIOLOGICAL PERIODIC TABLE[a,b]

H	
Li	F
Na	Cl
K	Br
Rb	I
Cs	At
Fr	117

Be	B	**C**	N	O
Mg	Al	**Si**	P	S
Ca	Ga	**Ge**	As	Se
Sr	In	**Sn**	Sb	Te
Ba	Tl	**Pb**	Bi	Po
Ra	113	**114**	115	116

Sc	Ti	V	Cr	**Mn**	**Fe**	Co	Ni	Cu	Zn
Y	Zr	Nb	Mo	**Tc**	**Ru**	Rh	Pd	Ag	Cd
La	Hf	Ta	W	**Re**	**Os**	Ir	Pt	Au	Hg
Ac	Rf	Ha	Sg	**Ns**	**Hs**	Mt	110	111	112

Ce	Pr	Nd	Pm	Sm	Eu	**Gd**	**Tb**	**Dy**	Ho	Er	Tm	Yb	Lu
Th	Pa	U	Np	Pu	Am	**Cm**	**Bk**	**Cf**	Es	Fm	Md	No	Lr

[a]Inert gases are excluded

[b]Elements in **boldface** are physically/chemically suited as a basis for living systems.

atomic-number radioactive elements formed in supernovae, has high nuclear stability, and therefore is anomalously abundant in the cosmos. Indeed, the periodic *table* can be reconfigured as a periodic *pyramid* for biological purposes.

Carbon is relatively stable to oxidation or reduction because of its central location in its electronic shell. Silicon would be an equally good basis for life, although somewhat rarer, but I can't help suspecting that computers are a form of artificial life (as opposed to artificial intelligence). I have seen the movie *Terminator*. In an oxidizing environment such as Earth's, elements on the right side of the pyramid appear to be relevant to life *as we know it*. Things may have been different prior to the advent of oxygenic photosynthesis.

Hydrogen was produced by the Big Bang and is the most abundant element in the universe. It too can be readily oxidized or reduced. It tends to be oxidized—that is, to lose its sole electron. This electron is responsible for electromagnetic flow within water, the biological solvent, and (indirectly) for hydrogen bonding and the solubility of ions and compounds in water. Water, or any fluid, also permits diffusion of matter, allowing chemical reactions. Thomas D. Brock of the University of Wisconsin, from whom I unfortunately never took a course, once mentioned that any protonated ionic salt (such as HBr) may serve to replace water as a biological matrix. This salt would, of course, have to exist as a fluid—gas, liquid, or possibly plasma. However, the nonbinding electrons of oxygen make H_2O the superior solvent. Alexander M. Klibanov of MIT, from whom I did once take a course, has demonstrated that biocatalysis can occur in nonaqueous environments. Life processes can occur within nonionic organic solvents, but only in the presence of some water. Thus, water is the biological solvent; it is the key to finding life in unusual environments. Apparently, my exposure to engineering (heat transfer and mass transfer) has paid off in my analysis of how life works.

Carbon compounds critical to life are present in the universe and were "rained down" on the nascent Earth. However, they did not "come alive" until they were able to exploit electronic flow. The proximal electron donor for life three to four billion years ago on prebiotic earth was iron, which forms the core of earth and other terrestrial planets. Transfer of electrons from iron within iron-sulfur compounds released small amounts of energy that powered the first membrane-bound assemblages of organic compounds—protocells. Thus cosmology and planetary science are critical to the origin of life. And iron—the stuff of which many planetary bodies are made—is a crucial ingredient.

As much as I love exploring the roles of carbon, hydrogen, oxygen, sulfur, and iron in the origin of life, acellular organic compounds containing these elements alone have not been shown to be capable of life. It is the presence of nitrogen—the "knowledge" element—that confers a shadowy form of life to viruses/viroids and prions. I suspect that it is the ability of nitrogen to form flat, stable aromatic compounds with carbon that is critical to nitrogen's importance; configuration is important to proper alignment of molecules. There must be some reason, in the context of biological receptors or ease of synthesis, that flat keys like those for houses are preferred to three-dimensional keys like those for bicycle locks. Additionally, I am heartened by the recent discovery that DNA, which is composed in part of aromatic nitrogen heterocycles, conducts electronic flow. And perhaps the slightly increased tendency of nitrogen toward oxidation, relative to carbon, is a good compromise with the significantly higher reactivity of oxygen.

Reflecting on the atomic—chemical/physical/cosmological—basis of life, I find myself considering how life forms interact, or the science of ecology. I immerse myself in the writings of Lynn Margulis. Dr. Allen first drew my attention to her work. I subscribe to the endosymbiotic theory on the origin of mitochondria and chloroplasts. I believe that endosymbiosis with nitrogen-fixing bacteria like *Rhizobium* has evolved to allow plants to exploit life's hunger for nitrogen. It is ironic that this gas predominates in Earth's current atmosphere. Meanwhile, I am captured by Woese's work. I would have sought to study with him had I gone to the University of Illinois. I almost did, but Wisconsin won. The theory that eukaryotes arose by endosymbiosis between bacteria and archaeans (*Thermoplasma* in particular) is tremendously heartening. The rise of different eukaryotic types by the mixing and matching of cell organelles, as described by Margulis, renews my faith in probability.

The rapid evolution of microorganisms—because of their short generation times—will produce a corollary to the Law of Microbial Infallibility. I believe that there exists at least one organism adapted to the degradation of any compound, even anthropogenic ones, found in any environment. I wonder what compound I will work on next. I am curious about the direction microbial ecology and evolution will take. I find it noteworthy that catabolic activity—biotic or abiotic—routinely generates carbon dioxide, a simple inorganic compound with a high degree of entropy. I am intrigued by the greenhouse effect and would love to see increased interest in the use of chemo- or photo-lithotrophic autotrophs for production of chemicals. But I also recognize that global warming through carbon dioxide accumulation may reflect the triumph of thermodynamics over life as we know it. Separately, I wonder if the preponderance of silicon dioxide in the lithosphere derives from another, silicon-based extinction event.

Now I see myself a teacher. My commitment is to help nontraditional students learn. My classroom is filled with them. Funded by NASA, my assignment is to train students for the new millennium—future interplanetary explorers perhaps—to recognize life on other planets. My fascination with mathematics, cosmology, physics, chemistry, and planetary science—and microbiology of course—has finally paid off. Oh, and my next stop perhaps will be Europa—maybe in 2010.

BRENDLYN D. FAISON was born in Lorain, Ohio, in 1954. She received her bachelor of arts degree in biological sciences from Wellesley College in 1975. She was a nondegree candidate in biochemistry at Columbia University from 1975 to 1977. During that time she also was a research laboratory technician at Albert Einstein College of Medicine. She received a master of science degree in applied biological sciences (food science and technology) in 1981 from MIT. She earned a doctoral degree in bacteriology from the University of Wisconsin in Madison in 1985. After receiving her Ph.D., she spent two years as a staff scientist at the Procter & Gamble Company and then worked as a staff scientist/research microbiologist in the Chemical Technology Division of the Oak Ridge National Laboratory. In 1996, while still employed by Oak Ridge, she was visiting Assistant Professor in Biological Sciences at Wellesley College. In 1998 she became Associate Professor at Hampton University. She has received awards of distinction and excellence for her published works from the Society for Technical Communication. Her research interests have been in industrial microbiology and bioprocess development and in evolutionary developmental microbiology. She serves as Chair of the Policy and Public Responsibility Committee of the Society for Industrial Microbiology and also as a director for that society.

The following papers are representative of her publications:

Hu, M. Z.-C., J. M. Norman, B. D. Faison, and M. E. Reeves. 1996. Biosorption of uranium by *Pseudomonas aeruginosa* strain CSU: Characterization and comparison studies. *Biotechnol. Bioeng.* **51:**237–247.

Faison, B. D. 1993. The chemistry of low-rank coals and its relationship to the biochemical mechanisms of coal transformations. In D. L. Crawford (ed.), *Microbial Transformations of Low-Rank Coals.* CRC Press, Boca Raton, Fla.

Faison, B. D. 1991. Biological coal conversions. *Crit. Rev. Biotechnol.* **11:**347–366.
Faison, B. D., T. M. Clark, S. N. Lewis, C. Y. Ma, D. M. Sharkey, and C. A. Woodward. 1991. Degradation of organic sulfur compounds by a coal-solubilizing fungus. *Appl. Biochem. Biotechnol.* **28/29:**237–251.
Faison, B. D., C. A. Cancel, S. N. Lewis, and H. I. Adler. 1990. Binding of strontium by *Micrococcus luteus*: Mechanism and potential applications. *Appl. Environ. Microbiol.* **56:**3649–3656.
Faison, B. D., T. K. Kirk, and R. L. Farrell. 1986. Role of veratryl alcohol in regulating ligninase activity in *Phanerochaete chrysosporium*. *Appl. Environ. Microbiol.* **52:**251–254.
Faison, B. D., and T. K. Kirk. 1983. Relationship between lignin degradation and production of reduced oxygen species by *Phanerochaete chrysosporium*. *Appl. Environ. Microbiol.* **46:**1140–1145.

Converting to Science

As Woese and others challenged the conventional wisdom that pervaded microbiology at the midpoint of the twentieth century, so had Lazarro Spallanzani attacked the religious and scientific belief in spontaneous generation that prevailed two centuries earlier. Born in 1729, just six years after the death of Antony van Leeuwenhoek, Spallanzani questioned the "known fact" that living microorganisms could arise from dead, decaying organic matter. Spallanzani, like van Leeuwenhoek, sought to expose the hidden things of nature. He studied at the University of Reggio to become a scientist and by age 30 was appointed to the faculty of that university. But he also became a Catholic priest, saying masses to help support his career as a scientist. It was the end of the Grand Inquisition, and questioning superstition and the conventional dogmas was both permissible and fashionable.

Spallanzani sought to answer the question: Does every living thing have parents? Even van Leeuwenhoek's lowly bacteria? Did God create every plant and animal in the first six days? And all those microbes too? How relevant these questions are even today with the renewed debates over the teaching of evolution. And who better than a priest to ask the fundamental question as to how life arose? As De

Kruif described Spallanzani: "Despising secretly all authority, he got himself snugly into the good graces of powerful authorities, so that he might work undisturbed. Ordained a priest, supposed to be a blind follower of the faith, he fell savagely to questioning everything, to taking nothing for granted—excepting the existence of God, of some sort of supreme being."

Truly exceptional was Spallanzani's ability to bridge religion and science, to move from subjective philosophies to objective experimental observations. By careful scientific observation he demonstrated that heat destroyed microbes and that spoilage was prevented indefinitely in sealed flasks. Showing that the decay of organic matter is caused by microorganisms that arise by the reproductive division of living microbes established that life begets life—which means microorganisms are like all other living organisms. In De Kruif's words: "[H]is genius whispered to him that the fantastic creatures of this new world were of some sure but yet unknown importance to their big brothers, the human species...."

Heroes, Saints, and Microbes

During my sophomore year at an all-boys Catholic high school in Kalamazoo, Michigan, I was introduced to the writings of Pierre Teilhard de Chardin. Teilhard de Chardin was a well-known archeologist, paleontologist, entomologist, and philosopher, but of greater importance to me was that he was also a Jesuit priest. He had decided by the age of 7 to enter the priesthood, but he was also drawn by the allure of science. It was in the mountains of his native France around Clermont-Ferrand that his visions of creation, including the origins of cells, began to take shape. His later writings synthesized his experiences on archaeologic digs of early hominoid remains in the deserts of China and Egypt with his intense prayer life. His books captivated me, particularly *The Phenomenon of Man* and *Hymn of the Universe*. He, unlike any other scientist with whom I was familiar, had found the way to merge his religious faith and his quest for scientific knowledge together to focus on the origins of man. Thus, based on the model of Teilhard de Chardin, I entered the University of Dayton aspiring to complete dual degrees in theology and biology and planning to enter religious life.

During my freshman year, Dr. Joseph J. Cooney, a marvelous and inspirational teacher, introduced me to the fascinating world of microbiology. I was captivated and decided that I wanted to pursue a career in biology. Yet theology still beckoned me. So I spent the summer after my freshman year in Europe studying theology in a tour of eight European universities, including my father's alma mater, the Free University of Amsterdam in The Netherlands. As my *raison d'être* began to emerge, it became clear that all the wonders and beauty of the various processes of life could be observed within the simplest of cells—bacterial cells—and my desire to study bacteriology took root.

By the end of my sophomore year, and in large part as a result of my studies in Europe, my educational plan began to change. A student research project in Dr. Cooney's laboratory on the pigments of *Micrococcus roseus* convinced me that I needed a stronger background in chemistry to complement my training as a microbiologist. Thus, some of my theology courses had to give way to make room for physical chemistry and more laboratory time. Also, my plans for entering a religious community during the summer were postponed after receiving an offer from Dr. David Taylor to work as a set-up technician in the clinical microbiology laboratory at St. Elizabeth's Medical Center in Dayton. I spent the summer as a microbiologist at St. Elizabeth's instead of as a novice in a religious community. Under the tutelage of Dr. Taylor, I was introduced to the world of clinical microbiology. Here I had a firsthand opportunity to see the world of microbiology come alive with astonishing relevance as physicians came to the clinical microbiology laboratory to learn about the infections in their patients and, more important, about which antimicrobial agents to select for therapy. It was here, after reading several dozen Kirby-Bauer disk diffusion plates, that my fascination with antimicrobial resistance began. Resistance clearly impacted therapy, and it was incumbent upon the clinical microbiologist to help the clinician interpret the information. Before long, it became clear that I had found my home in clinical microbiology.

In 1975, an alarming discovery was reported. A β-lactamase–encoding plasmid was isolated from an ampicillin-resistant strain of *Haemophilus influenzae*, a major cause of meningitis in children worldwide. In his book *Infectious Multiple Drug Resistance*, Stanley Falkow predicted that a similar plasmid would likely appear in the sexually transmitted organism *Neisseria gonorrhoeae* in the future. His prediction took only a year to be fulfilled. Thus, within barely a year's time, treatment of two very different epidemic infectious illnesses, bacterial meningitis and gonorrhea, were compromised by plasmid-mediated resistance genes. A better understanding of antimicrobial resistance mechanisms and the exchange of

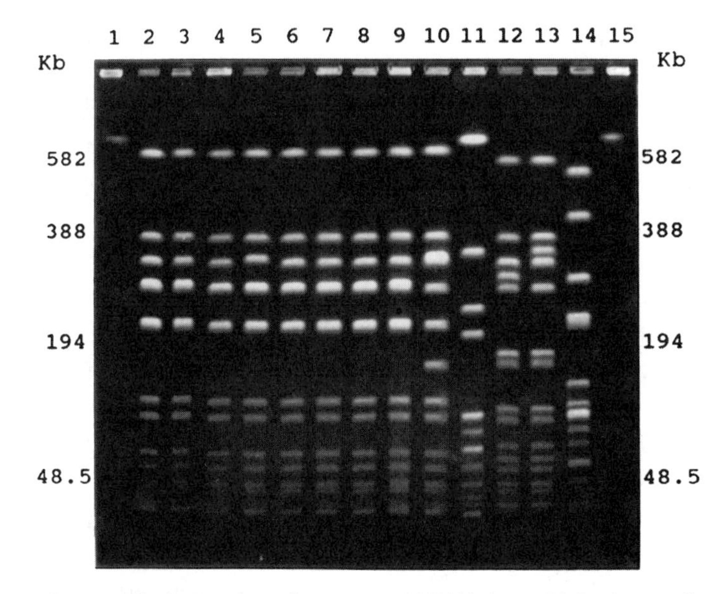

Agarose gel showing chromosomal DNA from 13 isolates of Staphylococcus aureus *after digestion with the restriction endonuclease* SmaI *and resolution using pulsed-field gel electrophoresis.*

genetic information among bacteria became a key research imperative. It was during this time that I entered the laboratory of Dr. Frank Young at the University of Rochester as a graduate student. After long discussions, we decided my project would focus on the genetics of antimicrobial resistance in *N. gonorrhoeae*, and I was excited about this opportunity to be a part of this evolving story. It was just the type of project—the evolution of microorganisms—that Teilhard de Chardin would have appreciated.

On the broader scientific scene, the era of restriction endonucleases and gene cloning was dawning and, with it, a growing fear in both scientific and public circles that these new genetic techniques would ultimately lead to cloning of humans and a variety of scientific atrocities. The ethical debate around the science of cloning (deemed at one point to be the "forbidden fruit" by Dr. Young, himself a religious preacher as well as scientist) was intense. Theologians and scientists engaged each other in active debate. I loved it!

As my graduate studies drew to a close, my long-term goal was still to direct a clinical microbiology laboratory. Thus, after completing my doctoral degree in Rochester, I headed west to Seattle to begin a postdoctoral fellowship in clinical microbiology and public health at the University of Washington. Because continuing my research on antimicrobial resistance was to be an integral part of my fellowship, Dr. Falkow was kind enough to provide laboratory space and support for me. I must admit I felt quite out of my league to be sharing space with the likes of Jim Kapur, Mike Koomey, Steve Moseley, Rodney Welsh, Alison Weiss, Dan Portnoy, and Pat Totten, but they were all very supportive. It was here that I developed my first DNA probe for a resistance gene, *aadB*, which encodes the ANT(2") adenylyltransferase, which in turn mediates resistance to gentamicin, kanamycin, and tobramycin. This gene was common in gram-negative bacilli at the Seattle Veterans Affairs (then called Veterans Administration) Medical Center, and the probe was instrumental in helping trace the spread of the gene as it disseminated through multiple bacterial species in the medical center, causing a series of multiresistant nosocomial infections. It was here that I was introduced to infection control issues.

During the "public health" portion of my training, I visited the Washington State Health Department, which at that time was located in the historic Smith Tower in downtown Seattle. One afternoon during a laboratory meeting, Dr. Jack Allard, then head of the laboratory, told us of a potential outbreak of *Yersinia enterocolitica* on nearby Vashon Island that seemed to be linked to contaminated tofu. As this was my first exposure to an outbreak investigation, I was fascinated, if not somewhat confused, by the discussions of "case-control" and "point-prevalence" studies, but felt more at home when the characteristics of the *Yersinia* strain were discussed. A dozen or so isolates were already available, but the serotyping would not be complete for several more days, leaving multiple questions about the suspected outbreak unanswered. That same evening I had invited several friends over for dinner. I was making a Szechuan meal that included hot and sour soup for which I had purchased tofu earlier in the week. I will never forget my shock when I went to my refrigerator and realized that the tofu I had purchased was from the same source identified in the outbreak investigation! Needless to say, I didn't put it in the soup, but I did take it to the lab the next day to culture it. Sure enough, *Y. enterocolitica* was isolated 24 hours later. But was it the outbreak strain? With a newfound sense of purpose, I asked Dr. Allard if I could have subcultures of the other isolates so that I could perform plasmid fingerprinting on them. From Dan Portnoy's work in Falkow's laboratory, I knew that *Y. enterocolitica* contained several plasmids, so fingerprinting seemed like a reasonable approach. Indeed, there were several different plasmid fingerprints observed among the many isolates they had collected, but more important, there appeared to be a common pattern among the "outbreak-related" strains. The strain from my tofu didn't match the outbreak pattern, but the fact that plasmid fingerprinting worked was joyous news. It was my first contribution to an outbreak investigation and the debut of molecular typing methods in the Washington State Health Department.

Plasmid fingerprinting now has given way to pulsed-field gel electrophoresis, a strain-typing tool based on the pattern of restriction endonuclease digestion products revealed after chromosomal DNA is cleaved with enzymes and separated on agarose gels using a novel electrophoresis chamber. Although the technique could be applied almost universally to most bacterial species, it was common at conferences to hear sometimes congenial, but at other times heated, discussions about how to interpret the results. Even among scientists at the Centers for Disease Control and Prevention (CDC), there was disagreement on what the "rules" for interpreting the results should be. Is any change indicative of a different strain, or could the positions of several bands change and still represent the "evolution" of the same strain? At a small conference in Park City, Utah, after two long days of staring at pictures and slides of many pulsed-field gels, the "genetic event" concept emerged. That is, a single genetic event (a mutation or an insertion or deletion of DNA into the chromosome) could cause one DNA fragment in the pattern to split into two smaller fragments, resulting in a three-band change on a pulsed-field gel (see figure on p. 39). Everyone around the table experienced the "Aha!" experience. We were convinced that this was the way to understand how the banding patterns "evolved" over time. Thus, the first interpretive criteria for pulsed-field gel electrophoresis—the results of a truly collaborative effort—were born. Although I have had the opportunity to work on many projects, this was perhaps the most satisfying. It wasn't a "religious" experience, but I'm sure Teilhard de Chardin would have been pleased with the efforts.

FRED C. TENOVER was born in 1954 in Kalamazoo, Michigan. As an undergraduate, he majored in biology and chemistry at the University of Dayton, where he received his bachelor of science degree in 1976. His graduate studies were in microbiology at the University of Rochester, where he received a master of science degree and a doctoral degree in 1980. He was a postdoctoral fellow in clinical microbiology and public health at the University of Washington from 1980 to 1982. He then was a consultant for the U.S. Agency for International Development through Boston University to the

Faculty of Medicine at Suez Canal University in Ismailia, Egypt, from 1982 to 1985. He also was an Adjunct Associate Professor in the Department of Laboratory Medicine and the Department of Microbiology in the School of Medicine at the University of Washington. From 1982 to 1990 he advanced from Associate Chief to Acting Chief of the Microbiology Section at the Veterans Affairs Medical Center in Seattle. During that time he also served as Chief of the Molecular Biology Section. In 1990, he moved to Atlanta to become Chief of the Antimicrobics Investigation Branch at the Centers for Disease Control and Prevention. He later became Chief of the Nosocomial Pathogens Laboratory Branch, a position he still holds. He also is Director of the Centers for Disease Control and Prevention/World Health Organization Collaborating Center for Global Antimicrobial Resistance Monitoring and Adjunct Professor of the Department of Pathology and Laboratory Medicine at Emory University School of Medicine and Adjunct Professor of Epidemiology at the Rollins School of Public Health of Emory University. He has won several awards and honors, including the James H. Nakano Citation in 1996 and the Public Health Service Outstanding Unit Citation in 1995. He is a fellow of the Infectious Diseases Society of America and of the American Academy of Microbiology and the recipient of the ASM's Becton Dickinson Award for Clinical Microbiology for the year 2000.

The following papers are representative of his publications:

Jorgensen, J. H., L. M. Weigel, M. J. Ferraro, J. M. Swenson, and F. C. Tenover. 1999. Activities of newer fluoroquinolones against *Streptococcus pneumoniae* clinical isolates including those with mutations in *gyrA*, *parC*, and *parE* loci. *Antimicrob. Agents Chemother.* **43:**329–334.

Smith, T., M. L. Pearson, K. R. Wilcox, C. Cruz, M. V. Lancaster, B. Robinson-Dunn, F. C. Tenover, M. J. Arduino, M. J. Zervos, J. M. Miller, J. D. Ban, and W. Jarvis. 1999. Emergence of vancomycin resistance in *Staphylococcus aureus*: Epidemiology and clinical significance. *N. Engl. J. Med.* **340:**493–501.

Tenover, F. C., M. V. Lancaster, B. C. Hill, C. D. Steward, S. A. Stocker, G. A. Hancock, C. M. O'Hara, N. C. Clark, and K. Hiramatsu. 1998. Characterization of staphylococci with reduced susceptibilities to vancomycin and other glycopeptides. *J. Clin. Microbiol.* **36:**1020–1027.

Levin, B. R., R. Antia, E. Berliner, P. Bloland, S. Bonhoeffer, M. Cohen, T. DeRouin, P. I. Fields, H. Jafari, D. Jernigan, M. Lipsitch, J. E. McGowan, Jr., P. Mead, M. Nowak, T. Porco, P. Sykora, L. Simonsen, J. Spitznagel, R. Tauxe, and F. Tenover. 1998. Resistance to antimicrobial chemotherapy: A prescription for research and action. *Am. J. Medical Sci.* **315:**87–94.

Tenover, F. C., R. D. Arbeit, R. V. Goering, P. A. Mickelsen, B. E. Murray, D. H. Persing, and B. Swaminathan. 1995. Criteria for interpreting pulsed-field gel electrophoresis patterns. *J. Clin. Microbiol.* **33:**2233–2239.

Tenover, F. C. 1988. DNA probes for infectious diseases. *Rev. Clin. Microbiol.* **1:**82–101.

6

CAROL NACY

Succumbing to the Mysteries of Microbiology and Immunology

I would love to tell you a tale of early science precocity, of interest in nature and the wonders of life from my earliest moments, but such a tale would be truly fiction. I was not especially drawn to science as a child, or even in high school, although I always enjoyed math. I was quite good at all aspects of school, however. My father's career as a U.S. Army officer required moving the family over a dozen times before I left for college, giving me ample opportunity to adapt and adjust to school systems throughout the nation. My parents were the stabilizing forces in our mobile family, and I dearly wanted to please them by excelling in a career they found compelling: medicine. My goal in college was to become a doctor, but along the way I discovered the mysteries of microbiology. I was hooked, and my conversion to microbiology and immunology followed.

My early undergraduate years at The Catholic University of America in Washington, D.C., seem a blur today, and not just because I have put a fair distance between then and now. Each year of matriculation was marked by an ever larger political and social event that made classroom learning pale in comparison: the historic university-wide strike for academic freedom that pitted Catholic University against the Vatican in the dismissal of the university's Father Curran for teaching about birth control; the city-wide racial riots and Martin Luther King Jr.'s D.C. Freedom March during the Civil Rights movement; the nationwide student protests against the war in Vietnam, and of course Woodstock. These were amazing times. In between and around these major societal upheavals, I studied biology, botany, anatomy, physiology, genetics, chemistry, physics, calculus—all courses designed to ease me into medical school—and I waited to be inspired.

At the beginning of my senior year, with high anxiety that I had embarked on the wrong career, I signed up for a graduate-level pathogenic microbiology course taught by Dr. Gene Kennedy. For the first time in three years, I was fascinated from the beginning to the end of a course. Second semester brought immunology into my life and I was entranced. I forgot medical school and I was on my way—but to where?

I had no earthly idea whether I could succeed in research, and research seemed to be the main focus of graduate students in microbiology. I was thrilled to land a job immediately after graduation at the National Institutes of Health (NIH) at a time when the "baby boomers" flooded the market with untrained biology majors by the thousands. I worked in the laboratories of Dr. John Sever at the National Institute for Neurological and Communicable Diseases and Stroke and there got a taste of research as we sought the viral etiology of certain congenital malformations. Parenthetically, Dr. Sever was the recipient of the 1996 American Society for Microbiology (ASM) Abbott Laboratories Award in Clinical and Diagnostic Immunology; I was thrilled to present this prestigious award to my first scientific mentor at the ASM General Meeting.

Two years at NIH convinced me that I had the temperament for research and also honed my interest in understanding how we respond to and resolve infections. In 1972, I left NIH and returned to Catholic University to study with Gene Kennedy and learn everything I could about microorganisms and host response to infection. Catholic University is a teaching institution, and the microbiology professors (Gene Kennedy, Dick DeCicco, and Ernie Cutchins) generously filled me with details and helped me learn the tricks of our trade. I am still amazed at the time and energy they expended on each of their students; they are my role models.

I loved graduate school and spent all my waking hours at the university. Although immunology was my passion, I diverted my attention several times to interesting projects in totally unrelated fields. Under Dick DeCicco's guidance, I wrote an NSF Student-Originated Studies grant and received funding to explore *Hydrogenomonas* (an autotroph with hydrogen as an energy source) as a food supplement. Leading ten undergraduates and early graduate students, we filled the department with huge fermenters of water and devised intricate methods of piping hydrogen in and salvaging bacterial mass out of each fermentation. To our amazement, mice fed the dried bacteria grew at the same rate as mice fed mouse chow. It surely did not look appetizing to us—gray and flaky, and definitely a marketing challenge. I do not think anyone has proposed dried bacteria as a commercially viable product, although it was great fun seeing bacteria grow in just water.

On long nights, many of the microbiology and the cell biology graduate students pooled resources and put on midnight dinners in the seminar room, complete with candlelight and music. We became adept at autoclave cookery, producing some of the best stews and soups by adjusting the autoclave pressure and experimenting with timing. We baked pies in the drying oven and distilled—well, we won't get into that. Suffice it to say, we feasted in style and helped each other through the usual ups and downs of graduate research. It was hard to leave that nurturing environment, but I eventually tore myself away by finishing my degree requirements.

My graduate research was on *Streptococcus pneumoniae*, specifically on the immunologic mechanisms behind a "ribosomal" vaccine, a field of research popular at that time (early 1970s). I isolated the immunologically active principle of this vaccine, which was not RNA at all. It was contaminating lipoteichoic acid, part of the pneumococcal cell wall. Nearly twenty years later, a colleague and I began working with lipoteichoic acids as potential antigens to protect against gram-positive bacterial septic shock at EntreMed, Inc. In all that time, no one reported using lipoteichoic acids as a vaccine. I discovered, however, that someone actually patented lipoteichoic acid–RNA complexes in 1980, which foiled our attempts to claim lipoteichoic acid as an antigen in our own patent. You need a crystal ball in this business; who knew that my "contaminant" might be important for an entirely different use?

S. pneumoniae was a little fastidious for my taste. I worked with type II *S. pneumoniae*, which managed to lose its capsule with the least provocation. I longed to work with *Escherichia coli*, which thrived under conditions even graduate students could produce. I have never worked with *E. coli* in all my career, however. It is undoubtedly my punishment for dastardly deeds of a former life that I am destined to work with fragile microorganisms.

Immediately after I finished my Ph.D., I began a National Academy of Sciences NRC Postdoctoral Research Associateship at the Walter

Reed Army Institute of Research (WRAIR) working on—rickettsia! I worked with Dr. Joe Osterman in the Department of Rickettsial Diseases on *Rickettsia tsutsugamushi*, obligate intracellular bacteria that die by the log for each fifteen minutes they remain outside of cells. My cell culture techniques improved; in fact they became (of necessity) very fast. I studied the interaction of this intracellular pathogen and macrophages, white blood cells now known to participate in both the afferent and efferent immune systems. Thus began a consuming interest in the role of macrophages in host defense to intracellular pathogens.

Among other interesting findings, I discovered that antibodies, the goal of every vaccine, actually helped rickettsia get into cells, where they successfully commandeered the host cell machinery for their own purposes (reproduction). Macrophages, which kill most bacteria they ingest, were defenseless against this intracellular pathogen under normal circumstances. By chance, I discovered that culture fluids from activated spleen cells could change the interaction of rickettsia and macrophages. These fluids enabled macrophages to kill the rickettsia, even in the absence of antibody. This discovery was pure heresy at the time. I struggled mightily to disprove the hypothesis, but all evidence I accumulated pointed to a nonantibody-mediated event.

The scientists at WRAIR were very focused on vaccines, so I sought help in understanding spleen cell–induced macrophage killing of rickettsia at the NIH, where I began a lifelong collaboration with Monte Meltzer, whom I married. Monte and I (and our many postdocs and colleagues) defined macrophage activation, a special state achieved by macrophages exposed to the protein products (cytokines) of antigen-activated lymphocytes that enable macrophages to kill a wide variety of intracellular and extracellular targets. Monte concentrated on extracellular targets (tumor cells, schistosomula, *Giardia*), while I attempted to understand the regulation of macrophage killing of intracellular pathogens (*R. tsutsugamushi*, *R. akari*).

As I look back on this extraordinary time, I am struck by how little we understood about the role of macrophages in immunity. These cells were characterized as the body's vacuum cleaner (macrophage = large eater) since the time of Metchnikoff at the turn of the century. Their function was to ingest, digest, and clean up the tissue in which they found themselves. Macrophages were not considered critical for antigen presentation yet, indeed the concept of antigen presentation was in its infancy, and they were certainly not known to secrete cytokines that could influence the direction and magnitude of antigen-specific immune reactions. Our studies on macrophage activation contributed to a growing body of knowledge that macrophages participate actively in developing specific immunity as initiators of immune reactions and as potent effector cells.

I moved to the Department of Immunology at WRAIR to continue my macrophage studies using a tiny protozoan parasite, *Leishmania major*, which uses the macrophage as its sole host cell in mammals. With this organism, my students, postdocs, and I extended observations on activated macrophages into control of infections in mice and charted the genetic control of macrophage intracellular killing, as well as cutaneous and systemic leishmanial disease. We determined which cytokines induced the two activated antimicrobial macrophage effector reactions—resistance to infection and intracellular killing—and which cytokines shut down the killing. And we discovered the effector molecule that activated macrophages use to kill intracellular pathogens—nitric oxide (Molecule of the Year in 1994).

The Army changed missions again, and Monte and I moved to a new Department of Cellular Immunology in Rockville, Maryland, where I began to study *Francisella tularensis*, a bacterial parasite of macrophages, and Monte began studies on HIV, a viral parasite of macrophages and lymphocytes. For three years, we compared and contrasted the interactions of macrophages with these pathogens with the pathogens we studied before. Many of the details were similar; more important, many were different. We made inroads in the understanding of *Francisella* infections and devised a polymerase chain reaction–based detection system and an attenuated and subunit vaccine, and we explored the use of immunomodulation with bacille-Calmette-Guérin (BCG) with great success. We could change the LD50 from 10^1 bacteria to 10^7 bacteria. These wide discrepancies allowed us to determine which cytokines were responsible for the extraordinary protection induced by BCG and correlate the activity of these cytokines in vivo with the production of nitric oxide by macrophages. The investigators then in my laboratory continue to publish on this interesting pathogen in laboratories at the Food and Drug Administration, the Red Cross, EntreMed, and elsewhere.

In 1993, I was given the opportunity to build a biotechnology company from the ground up. I left the government after seventeen years of productive research on macrophage to pathogen interactions and became the Executive Vice President and Chief Scientific Officer of EntreMed, Inc. It is a move I regret for not a minute. Several former colleagues joined me to establish state-of-the-art laboratories, and the company was a success in all objective parameters. We had major corporate partners for our most advanced technologies, a healthy early product pipeline, and in June 1996 we became a publicly traded company on the NASDAQ. Our mission was to discover innovative and under-appreciated technologies in academia or government laboratories and manage the development of these technologies into products for further clinical and commercial development by a corporate partner. We did this well, and we enjoyed learning the business end of medical research. My scientists worked on various projects, many of them outside the field of microbiology. Most of us were, however, microbiologists; our training in bacterial physiology, biochemistry, molecular biology, and animal modeling of diseases was broad enough to accommodate such diverse research areas as angiogenesis, cancer, and cardiovascular diseases.

In late 1996, I was asked by the NIH to serve as the immunologist on a panel to review the extramural program in tuberculosis (TB). What an awakening! As I sat through the presentations, I felt as though I had been asleep for lo these many years. Why did I not know that one-third of the world's population was infected with *Mycobacterium tuberculosis*? How did I miss that TB was the leading cause of death in women and was responsible for 26% of all avoidable deaths worldwide? The statistics for the global burden of TB were so frightening that it seemed impossible I had not heard the message before. The statistics struck right at my heart, and I did get the message this time. I also heard countless times that this vibrant research community, which had attempted to address the complete lack of modern-day tools (diagnostics, drugs, vaccines) for TB, was faced with an insurmountable brick wall—no interest by pharmaceutical companies. With growing concerns of multidrug-resistant TB, this lack of interest in commercial development of new tools for TB control portended a return to the untreatable and lethal TB of the nineteenth century.

As I sat at the table listening to the innovative approaches to TB diagnosis and treatment that were being explored in academic labs, I realized that there were few microbiologists with the same set of business experiences as I had gained through starting EntreMed. I resolved to find a way to help the TB research community move their basic research ideas out of the lab and into the clinic, where they might do the vast number of TB patients worldwide some good. As a result of this commitment, I left EntreMed in 1997 to found with some colleagues two very different entities, both focused on TB: the Sequella Global Tuberculosis Foundation, a nonprofit organization dedicated to assisting the TB research community in identification and early development of potential products for TB control, and Sequella, Inc., a for-profit company to commercialize new products (e.g., diagnostics, vaccines, drugs) for the treatment of TB. Both entities are doing well and are fulfilling their respective missions.

The foundation received a generous grant of $25 million in September 1999 from the Bill and Melinda Gates Foundation to organize a global initiative to develop a vaccine against adult pulmonary TB. The research will be transitional, not basic, and will concentrate on experiments in animals and humans that will illuminate the path to a successful global vaccine. We are seeking additional philanthropic funds to assist companies and academic investigators in the clinical development of new anti-TB drugs. The path is set for the foundation, and there is great excitement both within the foundation and the research community. Sequella, Inc., on the other hand, is working with investigators in universities around the world and has licensed some innovative, useful products for diagnosis and treatment of TB. It is currently raising money to fund the development of its products and hopes to have at least one of its new TB products in the marketplace by the year 2001.

I have no idea how long this business-of-science segment of my career will last, but I look forward to whatever new direction is ahead. I have been blessed with a rich and rewarding life: the love of my family, my five beautiful and talented children and brilliant husband; the friendship of my scientific colleagues; the nurturing of my professors and mentors; the passions and successes of my career. Don't let anyone tell you that you can't have it all; you can. I do. The only thing I have ever missed is sleep, and I can always sleep when I retire.

CAROL NACY was born in 1948 in Tokyo, Japan. She received her doctoral degree from The Catholic University of America in 1978. She was a research scientist at Walter Reed Army Institute of Research and Adjunct Associate Professor at Catholic University before joining the staff of EntreMed, Inc. Her research is in macrophage activation and immunology of parasitic infections. She was President of the American Society for Microbiology in 1996. She served as the Executive Vice President and Chief Scientific Officer through EntreMed's 1996 public offering. She currently serves as Chief Executive Officer of a company she founded, Sequella, Inc., which creates new tools to combat the global tuberculosis epidemic.

The following papers are representative of her publications:

Sim, B. K. L., M. O'Reilly, H. Liang, A. H. Fortier, W. He, J. W. Madsen, R. Lapcevich, and C. A. Nacy. 1997. Recombinant human Angiostatin™ protein: An antiangiogenic protein with inhibitory activity against experimental metastatic cancer. *Cancer Res.* **57:**1329–1334.

Dijkstra, J., G. M. Swartz, J. J. Raney, J. Aniagolu, L. Toro, C. A. Nacy, and S. J. Green. 1996. Interaction of anti-cholesterol antibodies with human lipoproteins. *J. Immunol.* **157:**2006–2013.

Robledo, S., D. Leiby, C. Nacy, L. Toro, L. Valderrama, and N. Saravia. 1996. Cytokine induction, differentiation antigen, and adhesion molecule expression in human mononuclear phagocytes infected with *Leishmania (viannia)*. *FASEB J.* **10:**1060–1067.

James, S. L., and C. Nacy. 1993. Effector functions of activated macrophages against parasites. *Curr. Opin. Immunol.* **5:**518–523.

Mock, B., J. Blackwell, J. Hilgers, M. Potter, and C. Nacy. 1993. Genetic-control of *Leishmania-major* infection in congenic, recombinant inbred and F2 populations of mice. *European J. Immunogen.* **20:**335–348.

Leiby, D. A., R. D. Schreiber, and C. A. Nacy. 1993. IFN-γ produced *in vivo* during the first two days is critical for resolution of murine *Leishmania major* infections. *Microb. Path.* **14:**495–500.

Fortier, A. H., T. Polsinelli, S. J. Green, and C. A. Nacy. 1992. Activation of macrophages to kill *Francisella tularensis*: Identification of effector cells, activation cytokines, and effector molecules. *Infect. Immun.* **60:**817–825.

Mellouk, S., S. J. Green, C. A. Nacy, and S. L. Hoffman. 1991. Destruction of *Plasmodium berghei* sporozoites by hepatocytes involves nitric oxides. *J. Immunol.* **146:**3971–3976.

Davis, C. E., M. Belosevic, M. S. Meltzer, and C. A. Nacy. 1988. Regulation of macrophage antimicrobial activities: Cooperation of lymphokines for induction of resistance to infection. *J. Immunol.* **141:**627–635.

Mock, B. A., A. H. Fortier, M. Potter, J. Blackwell, and C. A. Nacy. 1985. Genetic control of systemic *Leishmania major* infection: Identification of subline differences for susceptibility to disease. *Curr. Top. Microbiol. Immunol.* **122:**115–121.

Haverly, A. L., M. G. Pappas, R. R. Henry, and C. A. Nacy. 1983. *In vitro* macrophage antimicrobial activities and *in vivo* susceptibility to *Leishmania tropica* infection. *Adv. Exper. Med. Biol.* **162:**433–439.

JOAN WENNSTROM BENNETT

Feminism, Fungi, and Fungal Genetics

It's been said that we see ourselves as characters in a story. My own story began without scientific aspirations. In elementary school, I had only the vaguest notions of what a microbiologist might be, and had I known, it was not something I would have aspired to become. My girlhood ambitions were embarrassingly conventional. I planned to become a fifth grade teacher (because it had been my favorite grade), marry Mr. Right, and stay home to raise children (because that was what women did back in the 1950s). In retrospect, the first seeds of change were planted in a prosaic garden: a Girl Scout troop in Brooklyn, New York, under the guidance of a remarkable leader named Doris Engborg. She taught us to identify trees by the shape of their leaves, wild flowers by the form of their flowers, and, most important, where to find nature in the city. When I was fourteen, my family moved to a suburb and finding nature became much easier. Our house was adjacent to a county park, and many of the unstructured hours of my adolescence were spent in those woodlands. "Hands-on experience at the critical time," writes E. O. Wilson, "not systematic knowledge, is what counts in the making of a naturalist." Freed from even the simple constraints of Girl Scouting, my immersion was decidedly of this unsystematic kind. As much of my time was spent sitting under a tree reading a novel as collecting empty birds' nests or identifying flowering plants. Mostly I wandered around, just off the path, happy to be alone in the woods. I was particularly attracted to "oddball" plant life such as skunk cabbages, Indian pipes, and fungi. Once after a heavy rain, I dug up a clump of mushrooms, planted them in a wooden box in the basement, and watered them until they turned into black slime. My mother was tolerant if not exactly supportive, preferring this quiet mycoculture to loud rock 'n' roll.

That same year, my high school biology class provided my first opportunity to look through a microscope. The cliché was true: The microscope revealed a new world. Pond water was filled with stunning algae and darting protozoa. Magnified mildews were weird tubes topped with bizarre spores. I loved it, and I loved my teacher, Rachel Ferraro. The class altered my goals: I wanted to become a high school biology teacher like Mrs. Ferraro.

I went to college in the early 1960s. At that time, most science departments had all-male faculties. Professional women were often shunted into teaching jobs at colleges away from the centers of research. Upsala College, where I did my undergraduate work, was one of these teaching colleges, and women taught most of my biology courses. Only years later, after my "consciousness had been raised," did my good fortune register. My scout leader, my high school biology teacher, and almost all of my undergraduate biology professors were women. They were committed teachers with exacting standards and a passion for science. They tacitly demonstrated that marriage and family didn't have to be divorced from work and science.

The summer between my junior and senior years of college, Dorothy McMeekin, my botany professor, steered me into a National Science Foundation program for undergraduates. I worked in the Plant Breeding Department at Cornell University, and my project involved a forage crop called bird's foot trefoil. There was work in the field scoring plants for desirable agronomic characters, and there was work in the laboratory examining chromosomes for possible cytogenetic aberrations. At the end of the summer, my supervisor, Robert Seaney, took me aside. "You have a knack for this sort of thing," he said. "You ought to go to graduate school and become a plant geneticist like Barbara McClintock." I replied, "What is graduate school and who is Barbara McClintock?"

Graduate school, I soon learned, was a place where you could get a tuition waiver and a stipend to work for an advanced degree. It seemed too good to be true. Barbara McClintock, I learned, was a prominent cytogeneticist. She had been elected to the National Academy of Sciences and had been president of the Genetics Society of America back in the 1940s when it was a rarity for women to have that kind of visibility. (Many years later, when McClintock won the Nobel Prize in Medicine and Physiology, I felt a frisson of reflected glory. The person paternalistically assigned to me as a role model, long before I was savvy enough to pick my own, was among the most brilliant experimentalists of this century.)

The fall of my senior year in college, I applied to eight graduate schools, seven in small towns and the University of Chicago to see if I could get in. Chicago gave me the best fellowship. I entered their Botany Department a few weeks after graduation, planning on a career in cytogenetics.

In our life stories, places as well as people play important roles. For me, the University of Chicago was such a place. It changed my life. The University of Chicago is an institution with "an attitude," a distinguished academic history, and a sink-or-swim educational philosophy. It was assumed that you already knew a lot, that you would work very hard, that you would accomplish something important, and that you would then publish your findings. Both then and now those assumptions evoke a vacillating sense of being either anointed or inadequate. While at the University of Chicago, I became a geneticist but not yet a microbiologist. Under the mentorship of Edward Garber, I earned a master's degree focusing on the cytogenetics of a green plant and found out that I was not cut out for looking through a microscope all day. I then switched to a Ph.D. project on the genetics of a little-known mold called *Aspergillus heterothallicus*. Supported by a National Institutes of Health Training Grant on Genetics, my fellow graduate students all seemed to be working on *Escherichia coli* or one of its phages. Some of those students mocked my organism because filamentous fungi were so far from the mainstream of molecular genetics and because they took so long to grow (up to a week!) compared to bacteria and bacteriophages.

The president of the University of Chicago at that time was George Beadle, who, with Edward Tatum, had promulgated the one-gene, one-enzyme theory. Beadle and Tatum had done their groundbreaking research using the red bread mold *Neurospora crassa*. Although Beadle rarely visited the Botany Department, his proximity sent an implicit message: Some people may have thought that all molecular biology was done with bacteria and their phages. Others may have found fungi terminally boring and scientifically unfashionable. But it was okay to study molds. Research on fungi could lead you to the Nobel Prize and to the presidency of a major university.

Working on fungi in a Botany Department created another disciplinary dissonance. Fungi were supposed to be plants, but it was hard for me to understand why. For starters, fungi were not photosynthetic. They could not make their own food, nor did they engulf it. Their filamentous cells located nutrients by exploratory growth. They digested their way through the environment and then absorbed what they had digested. Although fungi had cell walls like plants, these walls were made of chitin, not cellulose.

As a graduate student I wasn't confident enough to make the paradigm leap and derive the obvious conclusion: Fungi were not plants. Bacteria and fungi had been studied in botany departments more out of historical precedent than out of sound taxonomy. Nowadays, modern taxonomists have placed fungi in their own kingdom, a kingdom of eukaryotes on equal footing with green plants and multicellular animals. Fungi span the boundary between microbiology and macrobiology. The yeast *Saccharomyces cerevisiae* is not only the best-known model system for fungal life, it is also one of the best-understood model organisms for eukaryotic life.

By the time I finished my Ph.D., I was married and did what many young brides do: I followed my husband to the place where he had found a good job—the city of New Orleans. Lucky for me, the Agricultural Research Service had a major regional facility there. The Southern Regional Research Laboratory conducted targeted research on economically important agricultural problems. I was awarded a postdoctoral fellowship to work on the aflatoxin problem.

Aflatoxins are carcinogenic metabolites produced by several filamentous fungi in the genus *Aspergillus*. Aflatoxin contamination of food crops is an international health hazard. The fungi that make aflatoxin lack sexual phases (mycologists call them "imperfect"), and in the days of pre-recombinant DNA, it was almost impossible to conduct genetic studies on imperfect fungi. However, a few imperfect fungi possessed an "alternative to sex" called the parasexual cycle. My job was to elucidate the parasexual cycle in *Aspergillus parasiticus*.

It was my debut as an independent scientist. My boss, Dr. Leo Goldblatt, gave me a laboratory, a supply budget, and a blessing. Then he left me alone. With few exceptions, my colleagues at the Southern Regional Lab were all chemists. They also left me alone. The University of Chicago training proved invaluable. Working by myself, I happily isolated mutants, studied secondary metabolism, elucidated the parasexual cycle, and learned a lot of organic chemistry. In search of some biologists to talk to, I joined the American Society for Microbiology (ASM), an organization that has played an important part in my life ever since. One of the first papers I presented was at a meeting of the South Central Branch of ASM in 1970. Bill McDonald, a bacterial geneticist at Tulane University, heard the paper. Later that year, after an unexpected faculty resignation left Bill desperate for someone to help teach genetics, he remembered my talk. In 1971, I was hired by the Biology Department at Tulane University and became the first woman in a tenure-track position in that department.

I was ecstatic. And I knew the rules—publish or perish. The parasexual cycle, however, was not going well. It was a slow and clumsy way to do genetics. If I stayed with it, I'd never get tenure. Luck was on my side again. Scientists at MIT had just initiated research on the biosynthesis of aflatoxin and shown that the chemical skeleton came from acetate units. Nothing else was known of the pathway. Following the example of George Beadle, who had pioneered the use of blocked mutants to elucidate biochemical pathways, my collection of blocked aflatoxin mutants became a valuable resource. In collaboration with Louise Lee, a chemist at Southern Regional, we used the mutants to dissect some of the early stages of aflatoxin biosynthesis. These experiments were like playing a complicated game, and we had fun doing our biosynthetic research. When we were done, we published our findings. Soon we were invited to speak at national meetings, which was fun, too. My climb up the academic ladder was uneventful, and I received tenure the same year my third child was born. I was promoted to Professor of Biology in 1982.

Meanwhile, molecular biology was undergoing a transformation. The recombinant DNA revolution made eukaryotic organisms accessible—and fashionable—again. An industrial mycologist friend, Linda

Lasure, and I were asked by Arny Demain to organize an ASM Conference on gene manipulations in fungi, which in turn led to several edited books on the topic. My group continued to study aflatoxin genetics and biosynthesis, benefiting from close collaborations with the Southern Regional Laboratory, first under the leadership of Alex Ciegler and later with Deepak Bhatnagar and Ed Cleveland. It has given me enormous satisfaction to see "my" blocked mutants applied by younger colleagues to develop aflatoxin biosynthesis as a model system in the molecular biology of secondary metabolism. More recently, with the help of Brendlyn Faison, my group at Tulane has branched out to apply fungal degradative metabolism to environmental problems, characterizing new species for bioremediation. I have also taken a strong interest in the genomics of filamentous fungi.

I've learned that being an academic scientist is a lot more than performing experiments in the laboratory. It involves collaborating with colleagues, writing grant proposals, organizing symposia, editing manuscripts, teaching courses, supervising graduate students, traveling to meetings, generating peer reviews, and a fair amount of "political" science. In many instances, diplomatic and social skills are as important in making a project work as scientific hypotheses and ideas. When my children were young I did a lot of scientific editing, because a manuscript is easier to bring home than an experiment. As the kids have gotten older, I have spent increasing time working with professional societies, particularly ASM, the Mycological Society of America, and the Society for Industrial Microbiology, all the while spreading the gospel of fungal metabolism. It's been a wonderful life.

Is there a moral to my biographical narrative? I like to think so. The moral is that you don't have to be a genius with a precocious childhood to have a successful scientific career. Nor do you have to give up a "normal" family life to pursue a nonconformist topic such as the sex life of obscure fungi. Each of us has a passion and a story. Each of us can find appropriate mentors and institutions. Abilities and ambitions aren't fixed at some point in early childhood; they change, and we change them, as we go along. My ardor for microbiology, feminism, and fungi is not appropriate for everyone, but it illustrates how in science, an ordinary but focused person can lead an extraordinary life.

JOAN WENNSTROM BENNETT was born in Brooklyn, New York, in 1942. She received a bachelor of science degree from Upsala College in 1963, completed her master's degree at the University of Chicago in 1964, and received a doctoral degree from that same institution in 1967. Upsala College granted her an honorary doctorate in 1990. Her research specialty is in fungal genetics and secondary metabolism. She is actively involved in advancing the roles of women in science and has served as Vice President of the British Mycological Society and as President of the American Society for Microbiology. She is currently Professor of Cell and Molecular Biology at Tulane University in New Orleans, Louisiana.

The following papers are representative of her publications:

Wunch, K. G., W. L. Alworth, and J. W. Bennett. 1999. Mineralization of benzo[a]pyrene by *Marasmiellus troyanus*, a mushroom isolated from a toxic waste site. *Microbiol. Res.* **154:**75–79.

Bentley, R., and J. W. Bennett. 1999. Constructing polyketides: From Collie to combinatorial biosynthesis. *Annu. Rev. Microbiol.* **53:**411–446.

Bennett, J. W. 1998. Mycotechnology: The role of fungi in biotechnology. *J. Biotechnol.* **66:**101–107.

Bennett, J. W., P. K. Chang, and D. Bhatnagar. 1997. One gene to whole pathway: The role of norsolorinic acid in aflatoxin research. *Adv. Appl. Microbiol.* **45:**1–15.

Kale, S. P., J. W. Cary, D. Bhatnagar, and J. W. Bennett. 1996. Characterization of experimentally induced, nonaflatoxigenic variant strains of *Aspergillus parasiticus*. *Appl. Environ. Microbiol.* **62:**3399–3404.

Bennett, J. W., and R. Bentley. 1989. What's in a name? Microbial secondary metabolism. *Adv. Appl. Microbiol.* **34:**1–28.

Bennett, J. W., and S. B. Christensen. 1983. New perspectives on aflatoxin biosynthesis. *Adv. Appl. Microbiol.* **29:**53–92.

Bennett, J. W., J. J. Dunn, and C. I. Goldsman. 1981. Influence of white light on production of aflatoxins and anthraquinones in *Aspergillus parasiticus*. *Appl. Environ. Microbiol.* **41:**488–491.

Bennett, J. W. 1979. Aflatoxins and anthraquinones from diploids of *Aspergillus parasiticus*. *J. Gen. Microbiol.* **113:**127–136.

CHEMICAL METAMORPHOSES

"No more shall spontaneous generation rear its ugly head." With this proclamation, Louis Pasteur declared the victory of science over myth. Considered by many the father of modern microbiology, Pasteur had been trained as a chemist. This training served him well as he pursued scientific questions that led him to explore the world of microorganisms—a world of unique metabolic activities and hitherto unknown physiologies. Pasteur would identify a problem, seek out all the available information, form a hypothesis, and devise experiments to test the validity of that hypothesis. Experimentation—like a chemist, rather than by natural observation like most of the biologists of the time—hallmarked the birth of microbiology and with it the modern era of biological science.

An aggressive, ambitious, argumentative, and highly patriotic Frenchman, Pasteur wanted to solve practical problems that would establish the economic and social superiority of France. Why wine spoiled and why French beer was inferior to German brews were the problems of nineteenth-century France that motivated Pasteur's quest for the scientific solutions that would reveal the power and diversity of those unseen microbes. Discovering that living microbes—and not just simple chemical reactions—are the real cause of the fermentations that change sugar to alcohol and wine to vinegar, and that cause milk to sour and food to spoil, transformed Pasteur the chemist into Pasteur the microbiologist. As De Kruif says, "he showed the world how important microbes were to it, and in doing this he made enemies and worshipers; his name filled the front pages of newspapers and he received challenges to duels; the public made vast jokes about his precious microbes.... in short it was here he hopped off in his flight to immortality...."

JOHN LYMAN INGRAHAM

Morphing Chemistry into Microbial Physiology

I had no doubts about chemistry until I started playing squash. That happened in 1947, but before it happened I had decided to earn a Ph.D. in chemistry at Stanford University. I was about to graduate with a bachelor of science degree from the College of Chemistry at the University of California at Berkeley after a stimulating undergraduate exposure to the field. World War II and service in the Pacific on the USS *Cowpens* CVL25 interrupted graduation, but that gave me, among other things, the G.I. Bill—no worry about tuition and living expenses. I knew chemistry was a fundamental science. It told you what everything was made of and how it changed.

Then squash came along. It was a wonderful game—a civilized game of racquetball played with a smaller, longer-handled racket. I joined the university squash ladder. You could challenge a name above and if you won the match, move up. I soon found my skill level, repetitively challenging and being challenged by the same people at about the middle of the ladder. One of my squash skill-level cohorts was a young Assistant Professor of Microbiology, Roger Y. Stanier. After our games he talked eloquently about bacteria and what they did. I was astounded to learn they did chemical things; for example, they convert nitrogen gas to ammonia, or synthesize specific compounds that make people ill. I also learned that we have a sort of chemical kinship with them—some bacteria need the same vitamins as humans.

I soon knew microbiology was for me. I earned a Ph.D. in microbiology with Stanier at Berkeley, studying what vitamins and amino acids the water mold *Allomyces* needs to grow, how this fungus metabolizes glucose, and how the bacterium *Pseudomonas fluorescens* breaks down benzoic acid and uses this toxic compound as a source of nutrients. My investigations and those of others in Stanier's laboratory led to a great understanding of microbial metabolism, especially the catabolic pathways used by *Pseudomonas* species for the utilization of many different compounds. Stainer later left Berkeley to join the Pasteur Institute in Paris, where he continued to pioneer studies on microbial metabolism.

After graduating from the University of California at Berkeley, I spent five years as a researcher at DuPont plus two more with the U.S. Department of Agriculture. I then joined the faculty of the University of California at Davis and never left. I pursued several lines of research that shared a common theme: using mutant microbial strains to probe microbial activity, a field once called *physiologic genetics*. With graduate students, postdoctoral fellows, and visiting professors, we used this approach to study several diverse aspects of microbial metabolism, including the malo-lactic fermentation of wine, fusel oil (a mixture of higher-molecular-weight alcohols) formation by yeasts, loss of function at low temperature, the pathways of pyrimidine nucleoside biosynthesis, and denitrification.

The study of loss of function at low temperature was a continuing project designed throughout my years to answer the question: Why do bacteria and other microorganisms stop growing at low temperature? Chemical reactions slow when cooled but they do not stop. One expects bacterial growth to slow based on chemical principles, but at a certain low temperature—the minimum temperature for growth—growth stops completely. Our studies showed that many things go wrong simultaneously at the minimum temperature of growth, and metabolism therefore stops completely. By isolating and studying mutant strains with an increased minimum temperature of growth (we called them *cold-sensitive mutants*), it was possible to determine what single change increased the minimum temperature of growth of one particular mutant strain. We isolated cold-sensitive mutants of

Escherichia coli that were unable to grow below 20°C (the minimum temperature for growth of wild type *E. coli* is 8°C). A pattern developed as illustrated by one set of mutants—those unable to grow below 20°C in the absence of the amino acid histidine but that exhibited normal low-temperature growth in its presence. In these mutants, biosynthesis of histidine was cold sensitive. The mutations causing this type of cold sensitivity lay in *his*G—the gene encoding the enzyme that catalyzes the first step of the pathway, the one sensitive to feedback inhibition by free histidine. The wild-type form of the enzyme becomes progressively more sensitive to feedback inhibition as incubation temperature is decreased, probably because of weakening of hydrophobic bonds (which occurs in all proteins at low temperature) that changed the enzyme's conformation, rendering it more sensitive. Mutant forms of the enzyme were more sensitive at all temperatures. At 20°C the enzyme became so sensitive that the intracellular pool of histidine was inadequate to support protein biosynthesis.

The phenomenon of change in regulatory responses of proteins with temperature proved to be a general one; however, one cannot predict whether regulation becomes more or less severe as temperature is lowered. Proteins in which regulation is more severe are susceptible to mutation, causing cold sensitivity. Conformation changes caused by weakening of hydrophobic bonds at low temperatures can also affect assembly of cellular organelles. For example, certain mutations in ribosomal proteins cause cold sensitivity because they prevent ribosomal assembly at low temperature. From the study of cold-sensitive mutants, we concluded that bacteria stop growing at low temperature because weakening of hydrophobic bonds causes conformational changes in proteins that preclude growth largely by distorting regulation or stopping assembly processes.

As a university professor, in addition to carrying out an active research program, I was a teacher. I taught microbiology courses at all levels: introductory, upper division, and graduate. My favorite was always the beginning course. I imagined that some students might be as fascinated as I was on first hearing what microbes do. In 1974, my major professor and old squash partner, Roger Stanier, offered me a new teaching opportunity to reach a broader range of students. He invited me to join him and Edward Adelberg in producing the fourth edition of *The Microbial World*, which was the premier textbook widely used in microbiology courses. I replaced Michael Doudoroff, who had recently died. I was flattered because I used this textbook in my own courses and admired it. I was also challenged to the limit in writing this book and delighted when the project was over—again able to do something else at night and on weekends—but I learned that there is something strangely satisfying, fulfilling, maybe even addicting about textbook writing. You can present a field you love to newcomers in the way you think it ought to be presented. I went on to write several other books on microbiology and microphysiology. My latest is with my daughter, a physician, and targets nonmicrobiology majors heading into the health sciences. Writing these textbooks and my classroom teaching have been extraordinary opportunities to reach students and guide them into the field of microbiology, especially emphasizing the importance of microbial physiology.

Everyone's career has high points. So did mine. Probably the best came in 1993, when I was elected president of the American Society for Microbiology (ASM). I truly admire ASM because of the opportunities it provides for professional exchange and because it is a democratic organization with nearly equal numbers of men and women. With over 40,000 members, the ASM is the largest biological science society. It publishes high-quality scientific journals and books. Unlike most scientific societies, the ASM is thoroughly democratic: An expressed interest in microbiology is the only prerequisite for membership. I joined as a student.

I sometimes ask myself whether I would choose microbiology if I were beginning a career today. I think I would. It still holds the fascination of rich complexity; there are so many different kinds of microorganisms (probably only small fractions of existing microbial species have yet been discovered), and they do so many different things. In my opinion, microbiology passes the rigorous practical test of opportunity. Microbiologists—including those like myself who choose to specialize in studying the physiologies of microorganisms—will be needed in the future. Infectious diseases that were once thought to be under complete control are again active threats to human wellbeing. Emerging infectious diseases must be investigated. New ways must be found to control established pathogens that are rapidly becoming resistant to known antibiotics. The future of biotechnology depends on the skills of microbiologists, as does bioremediation. In addition, of course, we need microbiologists to teach and train aspiring microbiologists.

JOHN LYMAN INGRAHAM was born on September 22, 1924, in Berkeley, California. He received his bachelor of science and doctoral degrees in microbiology from the University of California at Berkeley. He was a research scientist in microbiology for DuPont and chemist for the Western Regional Laboratory for the U.S. Department of Agriculture before joining the faculty of the University of California at Davis. He was President of the American Society for Microbiology. He is currently Emeritus Professor of Bacteriology at the University of California, Davis.

The following books and papers are representative of his publications:

Ingraham, J. L., and C. A. Ingraham. 2000. *Introduction to Microbiology* (2nd ed.). Brooks/Cole Publishing Co., Pacific Grove, Calif.

Hsu, D., C. Z. Yuan, J. Ingraham, and L. M. Shih. 1992. Diversity of cleavage patterns of *Salmonella* 23S-ribosomal-RNA. *J. Gen. Microbiol.* **138:**199–203.

Neidhardt, F. C., J. L. Ingraham, and M. Schaechter. 1990. *Physiology of the Bacterial Cell: A Molecular Approach*. Sinauer Associates, Sunderland, Mass.

Ingraham, J. L., O. Maaløe, and C. F. Neidhardt. 1983. *Growth of the Bacterial Cell*. Sinauer Associates, Sunderland, Mass.

Stanier, R. Y., E. A. Adelberg, and J. L. Ingraham. 1976. *The Microbial World* (4th ed.). Prentice-Hall, Englewood Cliffs, N.J.

Waleh, N. S., and J. L. Ingraham. 1976. Pyrimidine ribonucleoside monophosphokinase and the mode of RNA turnover in *Bacillus subtilis*. *Arch. Microbiol.* **110:**49–54.

Guerola, N., J. L. Ingraham, and E. Cerda-Olmedo. 1971. Induction of closely linked multiple mutations by nitrosoguanidine. *Nat. New Biol.* **230:**122–125.

Shaw, M. K., A. G. Marr, and J. L. Ingraham. 1971. Determination of the minimal temperature for growth of *Escherichia coli*. *J. Bacteriol.* **105:**683–684.

Hoffmann, B., and J. L. Ingraham. 1970. A cold-sensitive mutant of *Salmonella typhimurium* which requires tryptophan for growth at 20 degrees. *Biochim. Biophys. Acta* **201:**300–308.

Squires, C. K., and J. L. Ingraham. 1969. Mutant of *Escherichia coli* exhibiting a cold-sensitive phenotype for growth on lactose. *J. Bacteriol.* **97:**488–494.

O'Donovan, G. A., and J. L. Ingraham. 1965. Cold-sensitive mutants of *Escherichia coli* resulting from increased feedback inhibition. *Proc. Natl. Acad. Sci. USA* **54:**451–457.

From Party-Going Chemist to Worldly Microbiologist

As long as I can remember, I wanted to be a scientist—specifically a chemist. After finishing my Ph.D. training as an organic chemist at the University of Nottingham in the United Kingdom, working on the structure and synthesis of fungal polyketides, I spent three formative and productive postdoctoral years in the United States, first in the laboratory of Gilbert Stork at Columbia University and subsequently with Gene van Tamelen at the University of Wisconsin. In both instances my work concerned natural products, namely the synthesis of plant terpenes and alkaloids. Imbued with the innovation and energy of American science, I returned (with an American wife and child) to England in 1959 to the post of Lecturer in Organic Chemistry at the University of Manchester Institute of Technology, all set for a career as a natural product chemist. However, I became disenchanted with the purely chemical approach to natural products and increasingly interested in learning about how these complex compounds were produced and what their functions were in nature, which led me to biochemistry books and the intriguing subject of amino acid biosynthetic pathways. I recall being particularly interested in the biochemistry of shikimic acid and tryptophan. This interest became more than a minor diversion when a group of students asked me to give some informal lectures on biological chemistry, and we soon had an enjoyable rump course, not for credit, going in the department. The blind leading the unseeing!

In 1961, I met Milton Salton at a party, the first of several key events in my career that were the result of partying! Milton was working on cell wall structure and the differences between gram-negative and gram-positive bacteria, and in response to my expressed interest, he invited me to work in his laboratory in the Department of Bacteriology at the University Medical School. This seemed like fun, and I started to learn about bacterial cell walls by trying to obtain pure preparations of strange onion-like structures called *mesosomes* (later shown to be artifacts!). My brief sojourn in Milton's lab convinced me that microbiology was where I wanted to be, and I asked him for help in finding a position in the United States that would permit me to learn more. My suggestions were the labs of Bernard Davis or Charles Yanofsky, in keeping with an earlier fascination with the aromatic amino acids. In no time I had resigned my lectureship at Manchester and arrived in Boston—a scientific immigrant with wife and two children in tow—to work with Bernie. My first day in the Shattuck Laboratories was hardly auspicious. There was a seminar by Salvador Luria on some aspect of phage T4 genetics; I had never heard of T4, knew nothing of genetics, and could not understand Salva! What *was* this microbiology stuff? But in time, with a lot of patient help from a variety of people in the department, including Mahlon Hoagland, Elmer Pfefferkorn, Luigi Gorini, Martin Lubin, and of course Bernie, I assumed responsibility for research projects with students, including M. D. Bissell and Porter Anderson, who worked on resistance to spectinomycin. I was also teaching a laboratory course, fielding difficult questions from medical students by suggesting that we go to the library together to look up the answer. This approach has served me well ever since.

In the meantime, my research project on the biochemical mode of action of streptomycin was moving along, when at the next party, I learned about Wally Gilbert's work on the separation of *Escherichia coli* ribosomal subunits and the demonstration that transfer RNA bound to the large subunit. This was the early 1960s, the exhilarating time of the cracking of the genetic code, with the Ochoa and Nirenberg labs racing each other to identify the codon triplets for each amino acid. Pierre Spahr in Paul Doty's lab taught me how to make polyuridylic acid (poly U), which I then used to show that poly U–directed polyphenylalanine synthesis is inhibited by streptomycin when a 30S ribosomal subunit comes from a streptomycin-sensitive *E. coli* but not when the 30S subunit is from a resistant strain. This demonstration of the specificity of streptomycin action was very exciting, the only problem being that Bernie did not like the result, because he believed that streptomycin targeted the membrane, not the translation process.

About this time I met Jim Watson (also at a party) and told him about my work. At his invitation, I gave a talk at his evening seminar (at which he read a newspaper, as usual), and he affirmed that the results were significant and should be published. I mentioned that my studies ran counter to the dogma in Bernie's lab, but Jim said that he would speak with Bernie. A few days later Bernie invited me into his office and told me that he thought that I should write up the work for Jim to submit to the prestigious *Proceedings of the National Academy of Sciences* (*PNAS*), but that he (Bernie) could not put his name on the paper. I did not object! Then followed an extraordinarily heady collaboration between Wally Gilbert, Luigi Gorini, and myself, emanating from Wally's ingenious interpretation of Luigi's "funny" streptomycin-compensated auxotrophs. His proposal was that the antibiotic bound to the 30S ribosome subunit and caused the substitution of an amino acid at the mutant site. We did a lot of experiments on poly U–stimulated translation in vitro, using a variety of antibiotics and radioactively labeled amino acids; it was awesome to see the counts increase when an aminoglycoside was added. This led to the discovery of antibiotic-induced errors and the role of the 30S ribosomal subunit in the translation of messenger RNA into protein; once again Jim submitted the paper to *PNAS*. Working with Luigi taught me to appreciate the power of bacterial genetics, which was reinforced on meeting François Jacob, who was visiting Harvard to present a lecture series, at yet another party, this time at Jeanna Levinthal's house. When he invited me to work in his lab at the Pasteur Institute in Paris, how could I refuse? So with wife and three children, I prepared to cross the Atlantic once again.

However, before this voyage, I arranged to work in Gobind Khorana's laboratory at the University of Wisconsin in Madison to find out if there was any specificity in aminoglycoside-induced codon misreading. This was a very intensive month during which David Jones and I tested the effect of aminoglycosides on the translation of every polynucleotide in Gobind's lab. The results supported the notion that the antibiotics have restricted activity on codon-anticodon interactions.

I was extremely fortunate to spend my first year in François Jacob's *grenier* at the Institut Pasteur sharing a laboratory with Ethan Signer, who had recently isolated ϕ80 transducing particles in collaboration with Jon Beckwith. Ethan and others in the group (Maurice Hofnung, Luis Pereira da Silva, and of course *le patron*) gave me a good grounding in bacterial genetics, and I finally learned what phage was really about. My conversion to a microbiologist was completed with the discovery that toothpicks had more important roles than just the removal of pieces of *boeuf bourgignon* from between one's molars. (Incidentally, French toothpicks did not work very well, and visits to the U.K. or U.S. were required to replenish our supplies; these precious items were autoclaved and reused until they finally split from the treatment and had to be reluctantly discarded. But that's another story.)

I have often thought about my switch from organic chemistry to microbiology. It was a change in my career path that I do not regret at all. Studies of natural product chemistry are enhanced by working on the microbes that make this wondrously diverse collection of compounds, and looking at microbes with the benefit of a chemical perspective adds to their fascination: What are all these small molecules doing?

JULIAN DAVIES was born in Neath, South Wales, Great Britain, in 1932. He received his bachelor of science from the University of Nottingham in chemistry, physics, and math in 1963. His doctoral degree in organic chemistry was granted in 1956, also from the University of Nottingham. He did postdoctoral research in the laboratory of Gilbert Stork, Department of Chemistry, Columbia University, and in the laboratory of Eugene van Tamelen, Department of Chemistry of the University of Wisconsin. He has an honorary doctorate of medicine from the University of Zaragoza in Spain and an honorary doctorate from the University of Guelph in Canada. He was lecturer in organic chemistry at the University of Manchester Institute of Science and Technology from 1959 to 1962. He was an associate in bacteriology and immunology at Harvard Medical School from 1962 to 1967 and also at the Institut Pasteur from 1965 to 1967. He was appointed to the faculty of the Department of Biochemistry at the University of Wisconsin in 1967 and became Professor there in 1970, a position he retained for 10 years. From 1974 to 1975 he also was a Visiting Professor at the University of Geneva. In 1980 he moved to Biogen in Geneva, first as Research Director and then as President. He was an Adjunct Professor at the University of Geneva from 1981 to 1985. He was Head of the Microbial Engineering Unit at the Institut Pasteur from 1986 to 1992. In 1992 he moved to the University of British

Columbia as Professor and Head of Microbiology and Immunology. He also served as Director of the West-East Center for Microbial Diversity from 1993 to 1996. He was President and Chief Executive Officer of TerraGen Diversity, Inc., from 1996 to 1998 and then became Chief Scientific Officer. Since 1998, he has been Vice President of research at TerraGen Diversity and Professor Emeritus at the University of British Columbia. He has received numerous awards and distinctions, including Distinguished Teacher Award of the University of Wisconsin in 1978; Hoechst-Roussel Award, American Society for Microbiology, in 1986; Microbial Chemistry Medal, Kitasato Institute, in 1991; Thom Award, Society for Industrial Microbiology, in 1993; Canadian Society of Microbiology Boehringer-Mannheim Award in 1997; Scheele Award, Swedish Academy of Pharmaceutical Sciences, in 1997; and the Bristol-Myers Squibb Distinguished Achievement Award in Infectious Disease Research in 1999. He is a fellow of the International Institute of Biotechnology, the Royal Society of London, the American Academy of Microbiology, and the Royal Society of Canada. He was President of the American Society for Microbiology in 1999.

The following papers are representative of his publications:

Jimenez, A., and J. Davies. 1980. Expression of transposable antibiotic resistance element in *Saccharomyces*. *Nature* **287:**869–871.

Berg, D. E., J. Davies, B. Allet, and J.-D. Rochaix. 1975. Transposition of R factor genes to bacteriophage l. *Proc. Natl. Acad. Sci. USA* **72:**3628–3632.

Benveniste, R., and J. Davies. 1973. Aminoglycoside antibiotic-inactivating enzymes in actinomycetes similar to those present in clinical isolates of antibiotic-resistant bacteria. *Proc. Natl. Acad. Sci. USA* **70:**2276–2280.

Helser, T. L., J. E. Davies, and J. E. Dahlberg. 1972. Mechanism of kasugamycin resistance in *Escherichia coli. Nat. New Biol.* **235:**6–9.

Yamada, T., D. Tipper, and J. Davies. 1968. Enzymatic inactivation of streptomycin by R factor-resistant *Escherichia coli*. *Nature* **219:**288–291.

Davies, J., and F. Jacob. 1968. Genetic mapping of the regulator and operator genes of the *lac* operon. *J. Mol. Biol.* **36:**413–417.

Davies, J., and B. D. Davis. 1968. Misreading of ribonucleic acid code words induced by aminoglycoside antibiotics: The effect of drug concentration. *J. Biol. Chem.* **243:**3312–3316.

Davies, J., D. S. Jones, and H. G. Khorana. 1966. A further study of misreading of codons induced by streptomycin and neomycin using ribopolynucleotides containing two nucleotides in alternating sequence as templates. *J. Mol. Biol.* **18:**48–57.

Davies, J., W. Gilbert, and L. Gorini. 1964. Streptomycin, suppression, and the code. *Proc. Natl. Acad. Sci. USA* **51:**883–890.

Davies, J. E. 1964. Studies on the ribosomes of streptomycin-sensitive and resistant strains of *Escherichia coli. Proc. Natl. Acad. Sci. USA* **51:**659–664.

10

HOLGER W. JANNASCH

Adventures Discovering Microbes Changing the Planet

I will never forget that afternoon in January 1977 when I got a telephone call from our port office's radio operator. It was relayed to me from the mother ship of ALVIN, our institution's research submersible. Every day our research vessels have to call home from wherever they are to report on their well-being and scientific news if there is any. On this particular day, ALVIN had been diving to 2,600 m depth at the Galapagos Rift (about 200 miles north of the Galapagos Islands) to find signs of the predicted seawater circulation through the freshly formed oceanic crust and the emission of hot water near tectonic spreading centers.

The geologist, who was the lucky one to be on this dive, landed in the midst of a copious population of invertebrates. When I first spoke to him on the phone, I was full of doubts, I must admit, that I heard correctly. Oceanographers knew for a long time that the deep sea floor looks like the Sahara desert: miles and miles of bare sediment with few animals here and there in permanent darkness and near freezing temperatures. This is simply because little of the organic matter—the animals' food source—produced photosynthetically at the sea surface reaches the deep sea through the sedimentation of particles. But now I was told about masses of mussels, huge white clams, and tubeworms

6 feet long. Since they even brought some of these animals up to the surface, it must have been right. Most surprising was the high biomass of these animals, which clearly could not be living on that limited amount of photosynthetically produced organic matter. Without photosynthesis, what would there be to feed those massive populations?

Well, microbiologists know about chemolithoautotrophy, or, in short, chemosynthesis. In chemosynthesis, instead of using light energy, the inorganic carbon CO_2 is reduced to organic carbon by chemical energy obtained, for instance, from the oxidation of ammonia, hydrogen, or hydrogen sulfide. This was discovered a long time ago, but in the biological carbon cycle, was never considered to amount to much in the presence of photosynthesis. Could it be that these deep-sea animals living in permanent darkness developed a life support system based on chemosynthesis?

When I was told that, indeed, hydrogen sulfide was contained in the warm springs in high concentration, I went right back to the lab and wrote a proposal to study the possibility that bacterial chemosynthesis may represent the base of the food chain for the existence of the astounding biomass production at these deep-sea hydrothermal vents, as they came to be called. And lo and behold, two years later (it takes considerable time to get funded for and prepare diving programs with ALVIN), a biology cruise went to the Galapagos Rift. This first expedition began a series of most exciting cruises as new vent sites with many different animal populations were discovered. These expeditions continued, and many new forms of hitherto unknown microorganisms and bacteria-animal interrelationships have been observed and described.

The necessary cooperation with colleagues of other disciplines for such work has always attracted me. Beginning as a microbiologist among limnologists, I was fascinated by the metabolic diversity of bacteria that took care of the remineralization of nutrients in the different parts of lakes—oxic, anoxic, acidic, alkaline—and I needed to know the physiography and chemistry of water bodies. Without the physical and chemical oceanographers, we would never have found the hydrothermal vents and their new biological world in the deep sea. Our learning from these colleagues about geochemistry has been paramount for us in understanding and predicting the extent of microbial life in the extreme corners of our biosphere.

During evolution, higher forms of life became limited to just two major metabolic systems: photosynthesis of the green plants and the

In the Alvin diving submarine.

digestion and respiration of organic carbon by the animals. Although higher forms could not exist without the metabolic abilities of the "primitive" microorganisms, the primitive microorganisms themselves could certainly exist without plants and animals. Harvard's paleontologist Stephen Jay Gould said in a lecture on evolution at Woods Hole that the 3.5-billion-year-old microorganisms will also be the ultimate survivors on this planet. Pasteur said, "The microbes have the last word."

But back to the deep-sea hydrothermal vents. It never fails to amaze me how and why this coexistence between the metabolically versatile bacteria and the genetically and developmentally advanced marine invertebrates produced interrelationships that appear to maximize the production of biomass. In fact, the electron donor at the base of this so highly efficient food chain is a poison: H_2S. Furthermore, the inefficient mechanism of feeding by filtering planktonic animals on the quickly diluting bacterial suspensions in the vent plumes is "cleverly" improved by developing various symbiotic systems where the bacteria grow autolithotrophically within certain tissues of the vent invertebrates: clams and tube worms. In turn, these animals, through a specially adapted blood system, provide the microorganisms with everything they need, especially their source of energy, hydrogen sulfide, and oxygen, and CO_2 as their source of carbon. How the microbially produced organic matter gets distributed in the animals is being studied in many laboratories.

It is interesting that the detour via deep-sea studies was necessary to discover these novel types of symbioses between chemosynthetic bacteria and marine invertebrates. Because many marine clams are known to occur in anoxic coastal marine sediments, a search for their symbiotic existence with chemosynthetic bacteria was immediately done. Sure enough, there is a clam living profusely in the H_2S-containing shallow sediments of Buzzard's Bay, right near Woods Hole, operating on the same principle. In the meantime, many other invertebrates have been found to make use of this symbiosis—a whole new area of research.

Another novel type of microorganism was found at the deep-sea vents. In the late 1960s, Thomas Brock and, later, Karl O. Stetter discovered so-called hyperthermophilic microorganisms, bacteria that grow at temperatures between 80° and 100°C, at terrestrial and shallow marine hot springs. We were soon also able to isolate many of these "extremophiles" from the deep-sea "hot smokers" where the temperature gradients range from 2° to 360°C, most of them belonging to Woese's new domain Archaea. Today these isolates have an important role in biotechnology, where highly temperature-stable enzymes, mainly polymerases and proteases, are commercially produced.

The mere observation of tremendous productivity of organic carbon (the copious animal populations on the deep-sea floor) from hydrogen sulfide as the main electron donor or source of energy in the presence of free oxygen leads to the logical question: Can we use a similar system for getting rid of one of our most bothersome waste materials of all mining industries and major source of acid rain (hydrogen sulfide) and at the same time use it for the production of useful biomass? We began work on this problem and devised a continuous flow system where bacterial biomass was harvested from a reactor fed with a H_2S/seawater mixture. We demonstrated that the produced biomass could be used for feeding mussels in aquaculture. Also, this well-defined carbohydratious material may be a useful base material for fermentations to alcohols as synthetic fuels or for other industrial applications. The oil prices are still too low to interest the government in financing the necessary upscaling of the process.

It is easy to see why this type of microbiology fascinates us, for it combines a healthy and always exciting mix of interdisciplinary activities—both classical and modern microbiological approaches. And, for anything, I wouldn't miss those dives to the deep-sea floor.

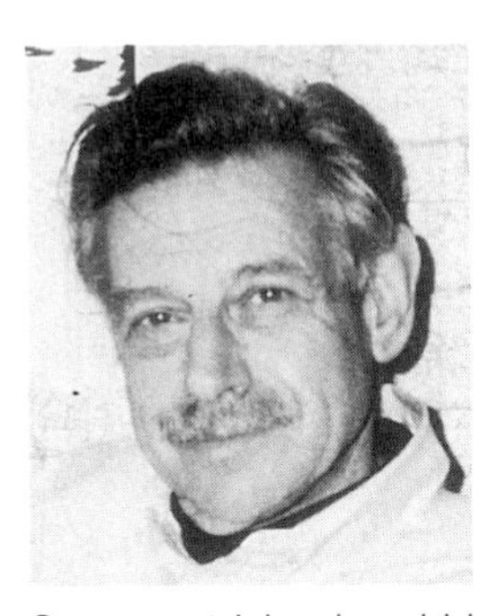

HOLGER W. JANNASCH was an internationally acclaimed marine microbiologist. He died in September 1998 at the age of 71. He was born, grew up, and went to school in the province of Silesia in Germany (now Poland). He received his doctoral degree at the University of Göttingen in 1955 and received subsequent appointments in the Max Planck Institute for Limnology, Plon, and at the University of Göttingen. He had an interest in and love for the oceans that came early, generated from his being a warden at a bird sanctuary on a small North Sea coast island and his between-semester work for several years on North Sea fishing vessels. During his student years, he demonstrated his adventurous nature by helping to found a limnological research facility in 1952 and obtaining a fellowship for a three-month stay in 1953 at the Stazione Zoologica in Naples, Italy, with which he continued his involvement throughout the years. A "microbiological prospecting" trip to Egypt in 1954 became the first of many travel adventures, which took him to all corners of the globe and combined microbiological science with a genuine appreciation for faraway lands and seas. In 1957 he joined Claude ZoBell's laboratory as a postdoctoral student in marine microbiology at the Scripps Institution of Oceanography. While in California he met C. B. van Niel, who had a major research and philosophical impact on the young microbiologist's developing career. He and the others who took van Niel's summer course in microbiology became in large part a continuing core of the Delft school of microbiology, passing this knowledge and philosophy on to future students. In 1963, he accepted an appointment as senior scientist at the Woods Hole Oceanographic Institution. He was attracted to the rather informal atmosphere of this American research institution and was not interested in "titles"; he wanted to be known throughout the scientific community simply as "Holger." He maintained close ties with his colleagues and friends in Europe, keeping himself abreast of and involved with developments in the European microbiological sciences. He originated and shaped the evolution of the microbiology course, now the microbial diversity course, at the Marine Biological Laboratory of Woods Hole. This course continues to teach a hands-on approach to the full range of microbes and their physiology and ecology and has influenced many career paths over the years. In 1977 Holger was the first microbiologist to be informed of the discovery of giant clams and tube worms surrounding deep-sea hydrothermal vents. In 1983 a newly isolated thermophilic methanogen was named in his honor, *Methanococcus jannaschii*. In 1996 the genome of this organism became the first archaeal genome to be completely sequenced. Holger was the recipient of the Henry Bryant Bigelow Medal in Oceanography from the Woods Hole Oceanographic Institution in 1980, the American Society for

Microbiology Fisher Scientific Award in Applied and Environmental Microbiology in 1982, the Cody Award in Ocean Sciences from the Scripps Institution of Oceanography in 1992, and the Fulbright Distinguished Scholar Award in 1992. In 1995 he was elected to the National Academy of Sciences as a Foreign Associate, and in 1996 the Holger W. Jannasch Chair was established at the Woods Hole Oceanographic Institution. His studies on the physiology and ecology of freshwater and marine bacteria, deep-water microbiology, and the growth of microorganisms at extreme temperatures and pressures, such as in deep-sea hydrothermal vents, have greatly advanced our understanding of microbial physiology and the extreme conditions under which some microorganisms grow.

The following papers are representative of his publications:

Teske, A., M. L. Sogin, L. P. Nielsen, and H. W. Jannasch. 1999. Phylogenetic relationships of a large marine *Beggiatoa*. *Syst. Appl. Microbiol.* **22:**39–44.

Blochl, E., R. Rachel, S. Burggraf, D. Hafenbradl, H. W. Jannasch, and K. O. Stetter. 1997. *Pyrolobus fumarii*, gen. and sp. nov., represents a novel group of archaea, extending the upper temperature limit for life to 113 degrees C. *Extremophiles* **1:**14–21.

Jannasch, H. W. 1997. Small is powerful: Recollections of a microbiologist and oceanographer. *Annu. Rev. Microbiol.* **51:**1–45.

Muyzer, G., A. Teske, C. O. Wirsen, and H. W. Jannasch. 1995. Phylogenetic relationships of *Thiomicrospira* species and their identification in deep-sea hydrothermal vent samples by denaturing gradient gel electrophoresis of 16S rDNA fragments. *Arch. Microbiol.* **164:**165–172.

Rueter, P., R. Rabus, H. Wilkes, F. Aeckersberg, F. A. Rainey, H. W. Jannasch, and F. Widdel. 1994. Anaerobic oxidation of hydrocarbons in crude oil by new types of sulphate-reducing bacteria. *Nature* **372:**455–458.

Jannasch, H. W., and T. Egli. 1993. Microbial growth kinetics: A historical perspective. *Antonie van Leeuwenhoek* **63:**213–224.

Gokce, N., T. C. Hollocher, D. A. Bazylinski, and H. W. Jannasch. 1989. Thermophilic *Bacillus* sp. that shows the denitrification phenotype of *Pseudomonas aeruginosa*. *Appl. Environ. Microbiol.* **55:**1023–1025.

Jannasch, H. W., and C. D. Taylor. 1984. Deep-sea microbiology. *Annu. Rev. Microbiol.* **38:**487–514.

Ruby, E. G., and H. W. Jannasch. 1982. Physiological characteristics of *Thiomicrospira* sp. Strain L-12 isolated from deep-sea hydrothermal vents. *J. Bacteriol.* **149:**161–165.

Schlegel, H. G., and H. W. Jannasch. 1979. Enrichment cultures. *Annu. Rev. Microbiol.* **21:**49–70.

Jannasch, H. W., and C. O. Wirsen. 1977. Microbial life in the deep sea. *Sci. Am.* **236:**42–52.

Jannasch, H. W., K. Eimhjellen, C. O. Wirsen, and A. Farmanfarmaian. 1971. Microbial degradation of organic matter in the deep sea. *Science* **171:**672–675.

Jannasch, H. W. 1969. Estimations of bacterial growth rates in natural waters. *J. Bacteriol.* **99:**156–160.

IN SEARCH OF A CAUSE

While Pasteur was saving France's fermentation industries—and occasionally pursuing the travails of human diseases—in Germany Robert Koch was studying to be a physician at the University of Göttingen. But when he began his practice, Koch became restless. Like Pasteur, he began searching for scientific answers to practical problems. According to De Kruif, Koch bemoaned the nonscientific state of medicine: "I hate this bluff that my medical practice is...it isn't because I do not *want* to save babies from diphtheria...but mothers come to me crying—asking me to save their babies—and what can I do?—grope...fumble...reassure them when I know there is no hope.... How can I cure diphtheria when I do not even know what causes it, when the wisest doctor in Germany doesn't know?"

Koch was right. As René Dubos wrote in *Mirage of Health*, "Until late in the nineteenth century, disease had been regarded as resulting from a lack of harmony between the sick person and his environment; as an upset of the proper balance between the yin and the yang, according to the Chinese, or among the four humors, according to Hippocrates. Louis Pasteur, Robert Koch, and their followers took a far simpler and more direct view of the problem. They showed by laboratory experiments that disease could be produced at will by the mere artifice of introducing a single specific factor—a virulent microorganism—into a healthy animal."

Prodded by his wife, Koch began to seek the answers, changing the view of the causes of infectious diseases and fathering medical microbiology and biomedical research as he did, and competing with Pasteur every step of the way for preeminence.

As a country physician, Koch began his scientific escapades isolated from any contact with the scientific community, working alone with primitive tools and materials. Over years of work, he and his colleagues developed the tools and methods that are still basic to the study of microbiology today—Petri dishes, agar media, Gram stains, and so forth. Most important, Koch demonstrated for the first time that germs grown outside the body would cause disease, and that specific microorganisms caused specific diseases. This was the beginning of Koch's illustrious career. It was a critical turning point for microbiology.

As Dubos wrote, "Koch and Pasteur wanted to show that microorganisms could cause certain manifestations of disease. Their genius was to devise experimental situations that lent themselves to an unequivocal illustration of their hypothesis—situations in which it was sufficient to bring the host and the parasite together to reproduce the disease. By trial and error, they selected the species of animals, the dose of infectious agent, and the route of inoculation, which permitted the infection to evolve without fail into progressive disease."

Koch established the means of discovering which microbes caused which infectious diseases. He set the path for diagnosing and eventually finding cures for those diseases. He wrote in 1884 in *The Etiology of Tuberculosis*, "To obtain a complete proof of a causal relationship, rather than mere coexistence of a disease and a parasite, a complete sequence of proofs is necessary. This can only be accomplished by removing the parasites from the host, freeing them of all tissue elements to which a disease-inducing effect could be ascribed, and by introducing these isolated parasites into a healthy animal with the resulting reproduction of the disease with all its characteristic features. An example will clarify this type of approach. When one examines the blood of an animal that has died of anthrax one consistently observes countless colorless, non-motile, rod-like structures.... When minute amounts of blood containing such rods were injected into normal animals, these consistently died of anthrax, and their blood in turn contained rods. This demonstration did not prove that the injection of the rods transmitted the disease because all other elements of the blood were also injected. To prove that the bacilli, rather than other components of blood produce anthrax, the bacilli must be isolated from the blood and injected alone. This isolation can be achieved by serial cultivation.... The serial transfers can be continued for 3 or as many as 50 passages and in this manner the other blood components can be eliminated with certainty. Such pure bacilli produce fatal anthrax soon after injection into a healthy animal, and the course of the disease is the same as if produced with fresh anthrax blood or as in naturally occurring anthrax. These facts proved that anthrax bacilli are the unique cause of the disease."

11

LUCY S. TOMPKINS

The Circuitous Path to Becoming a Physician-Scientist

I suppose my dream of becoming a physician began at age 5 or so with the recognition that our family pediatrician was an important part of my life, treating the usual pediatric infectious diseases, initially without benefit of antibiotic therapy. I set off to college at Stanford University with this dream, albeit I was somewhat ambivalent. It was the late 1950s, when a woman's place was in the home, or the classroom, or as a nurse, but certainly not as a doctor. Unfortunately, that view also prevailed in college, and I quickly became discouraged by numerous professors and by the prevailing chilly climate for women in medicine. So, I became a biology major, not a premed student, although I didn't really have a clear idea of how I would proceed in biology after graduation. I did think I might pursue a graduate program in immunology, stimulated by a wonderful professor at Stanford.

The first bend in the career path occurred when I married at a very young age and had to transfer to the University of Denver for my last year of college. After working as a research technician for 5 years, my chance came to do graduate work when, at the height of the Vietnam war, we moved to Washington, D.C. I applied to Georgetown University with the idea of doing an immunology Ph.D. program in the

Department of Microbiology. However, after admission, I learned that the immunology track was no longer open to new graduate students. After a couple of false starts, I approached a new professor in the department, Stanley Falkow, who was a pioneer in the "new" molecular biology of plasmids that carried antibiotic resistance genes. I was accepted as his first woman student. This was clearly a turning point in both my career and my life.

Although the tools we had at the time were primitive, we were working at the "cutting edge" of plasmid research and molecular biology, pioneering new ground for finding the causes of disease. We extracted DNA from cultures by precipitating it and winding it around a glass rod. We used a Model "E" ultracentrifuge to visualize plasmid DNA; there were no agarose gels. We analyzed replication of plasmids by radiolabeling the DNA and separating single- and double-stranded molecules by sucrose gradient centrifugation. We tirelessly examined the taxonomic relationship of one molecule or one species of bacteria to another by examining the hybridization kinetics. As interesting as my thesis work on plasmid replication was, I was especially intrigued by the mechanisms by which pathogens, like *Salmonella*, *Shigella*, and *Mycobacterium tuberculosis*, caused disease. My almost-buried primal dream of a medical career began to resurface, and I began to think that a career in infectious diseases might be a perfect marriage of microbiology and medicine.

At the end of my graduate program, I applied and was accepted at age 30 to Dartmouth's new three-year M.D. program, the oldest student ever accepted there. And I had a school-age son, also a unique situation for a Dartmouth medical student. This was such a novelty that the local newspaper covered my matriculation. By then, my career path was set for infectious disease medicine.

After completing my internal medicine residency at Dartmouth, I entered the fellowship program in infectious diseases at the University of Washington in Seattle. As part of the training, each fellow spent several weeks in the clinical microbiology laboratory, working at the bench alongside the technologists. I loved the diagnostic laboratory. Every day brings new diagnostic challenges, new "data," just as in the research lab, and moreover, it is a great place to explore the biology of infectious diseases.

I was also influenced greatly by another role model, Jim Plorde, the director of the diagnostic laboratory at the Veterans Administration

Hospital (VAH) and an infectious disease clinician. He conducted "plate rounds" in the clinical laboratory for the infectious diseases consulting team each morning in which we looked at smears, examined culture plates, and discussed clinical cases, essentially working out the medical mysteries of each case from the medical microbiology and clinical infectious diseases vantage points. This experience produced another turn in the road by stimulating me to combine my infectious disease fellowship training with a clinical microbiology fellowship at the University of Washington, with the ultimate goal of becoming a clinical microbiology laboratory director and an infectious disease clinician.

As part of the training program, I elected to do my research project in the Falkow laboratory, which had moved from Georgetown to the University of Washington in the early 1970s. At that point, recombinant DNA techniques, gene sequencing, cloning, agarose gel electrophoresis, endonuclease restriction enzyme analysis, and Southern hybridization were all being used in the lab. My project was to study the mechanisms of multiple antibiotic resistance of *Serratia* strains that had been isolated from patients at the VAH. During the course of analyzing these isolates using the new "modern" molecular techniques of plasmid agarose gel electrophoresis and restriction endonuclease analysis, I discovered that most of the patients were infected with a strain of *Serratia marcescens* containing a large plasmid that encoded many antibiotic resistance genes, including those that inactivated aminoglycosides. When I examined other *Serratia* isolates, which came from other patients at the VAH and from other geographical areas, it became clear that the infected patient strain had a unique plasmid profile that separated it from other strains, each of which also had a unique profile. It seemed that we could distinguish the strains isolated from patients who were epidemiologically linked (i.e., on the same unit, shared the same medical personnel, etc.) from other strains.

We also discovered that the large resistance plasmid carried by the *Serratia* outbreak strain was also present in several other enteric bacterial species isolated from patients in the same units. Using restriction endonuclease analysis, we showed that the plasmids in all of these antibiotic-resistant gram-negative bacilli were identical. In our published report, we used the term *molecular epidemiology* to describe the new technique to fingerprint bacterial strains. The

Seattle VAH plasmid, pLST1000, was later shown by others to be carried by other enteric bacterial strains isolated in many sites around the world, suggesting that plasmids can also be disseminated widely and appear under the influence of selective antibiotic pressure. We proposed that plasmid profile analysis, and possibly other molecular techniques, could be used in the clinical laboratory to differentiate strains of bacterial pathogens isolated from patients and reservoirs of infection.

The next opportunity to apply my ideas about the use of molecular fingerprinting came after moving to Stanford as the Medical Director of the Clinical Microbiology Laboratory. Shortly after my arrival, I learned that patients on the cardiovascular surgery service were developing very peculiar, even unique, *Legionella* infections, including prosthetic heart valve infections and other extrapulmonary infections that had rarely been described. Moreover, half of the infections were caused by *L. dumoffii*, which had rarely been seen as a cause of infection in humans. The other half were caused by *L. pneumophila* serogroup I, the major cause of Legionnaire's disease. Another strange aspect of the situation was that many of the patients had none of the usual risk factors; they had simply undergone elective cardiovascular surgery. We quickly discovered that the potable hot water in the hospital contained *Legionella pneumophila*, but we could not isolate *L. dumoffii* from any water source we tested. The question then was whether the *L. pneumophila* water strain was the same as the strain(s) in the patients. Because it was already known that there were many different strains of *L. pneumophila* serogroup I, we couldn't be certain whether the water strain(s) was the same as the patient isolates. Therefore, my colleagues and I in the Clinical Microbiology Laboratory applied a new molecular technique called *endonuclease restriction analysis* that would show whether the water strain and the patient isolates were the same. If so, we could presume that the water might be the reservoir of the infections and that transmission had occurred through some kind of water contact.

In conjunction with colleagues at the Centers for Disease Control and Prevention, we compared monoclonal antibody serotype results with restriction endonuclease analysis and showed clearly that the molecular fingerprint of the Stanford *L. pneumophila* water isolate was identical to all the patient isolates. Using the same techniques, we found that all the *L. dumoffii* patient isolates were also identical, suggesting that the potable water was a common source for both species. After six years and three major epidemiologic investigations, our group at Stanford finally confirmed the water as the reservoir of infection and discovered a mode of transmission from the potable water to the patients—through a sponge bath using the hospital water that was given to the patient immediately after surgery. Presumably, contaminated water containing *Legionella* sp. entered surgical wound sites around the mediastinal and pleural tubes used to drain fluids after surgery. We also finally discovered that *L. dumoffii* was also in the water, but only in water from two adjoining rooms in the intensive care unit (ICU) that we had not tested before. These rooms were used infrequently; thus, when the faucet was first opened to give the patient the sponge bath, it released a large bolus of *L. dumoffii* from the build-up of biofilm in the pipes. Having discovered that the hospital water was indeed the reservoir and that topical exposure to potable water was a route of transmission, we installed a continuous chlorination system for the water supply and discontinued the use of potable water in the ICU. No further nosocomial infections have occurred since 1989.

Solving this medical and epidemiologic mystery greatly influenced my subsequent career path in two ways. First, I began to realize my love of epidemiology, became the Director of the Hospital Epidemiology and Infection Control Program, and integrated molecular fingerprinting technology as a routine part of the program. We now have an entire section of the Clinical Microbiology Laboratory that performs molecular fingerprinting on an ongoing basis. Second, I also became enchanted with *Legionella* and have studied its molecular, genetic, and cellular basis of pathogenicity in my research laboratory ever since my first encounter with it at Stanford in 1983.

Perhaps the most exciting discovery of my career resulted from an encounter with a patient for whom I was asked to consult as an infectious disease physician. Mrs. TF was an immunosuppressed heart transplant recipient who had sustained a severe cat scratch six weeks before her admission to the hospital for evaluation of a prolonged febrile illness of unknown etiology. A computed tomography scan was performed that showed diffuse liver and spleen nodules, which were then biopsied and sent to the pathology lab. On reviewing the histopathology, we were astonished to see that the nodules were composed of new blood vessels and inflammatory cells, closely resembling a new condi-

Inoculating cultures in the clinical laboratory.

tion called *bacillary angiomatosis*, which had been recently recognized in patients with AIDS. When Mrs. TF's tissue was stained with the Warthin-Starry stain, it revealed myriads of tiny bacilli identical to those that had been seen in the skin lesions of patients with bacillary angiomatosis. These bacilli also resembled the bacteria seen in the lymph nodes of patients with cat scratch disease. This common syndrome was first described many decades ago but the etiologic agent was still a mystery.

In spite of our attempts, all the cultures from the tissues and blood of Mrs. TF were negative, leaving us puzzled about the nature of her illness. At this point, we turned to Stan Falkow, now at Stanford, and his postdoctoral student, David Relman, for advice on how to approach the diagnosis of an uncultured bacterial pathogen. We applied two new technologies, polymerase chain reaction and so-called universal 16S recombinant DNA primers, in a novel way to amplify the bacterial DNA from Mrs. TF's tissues and discovered that the sequence analysis revealed that her organism, and tissues from other patients with AIDS-associated bacillary angiomatosis, was closely related to *Bartonella (Rochalimaea) quintana*. This organism caused thousands of cases of trench fever in soldiers during World

War I but had rarely been seen in the United States. Subsequent studies showed that the new organism, now named *Bartonella henselae*, also causes typical cat scratch disease. Thus, these molecular and clinical investigations defined the nature of a new pathogen and revealed an especially intriguing aspect about *B. henselae* pathogenicity. On one hand, it has the propensity to cause a localized infection with regional lymphadenopathy, without angiomatosis, in people with normal immune systems, while on the other hand, it may produce an entirely different, and unique, pathology in those who are severely immunocompromised, especially patients with far-advanced HIV infection.

It is the dream of every microbiologist and infectious diseases clinician to discover a new disease or a new pathogen, but very few of us have this opportunity. Mrs. TF provided us with the good fortune to solve one old and one new medical mystery, and this serendipitous case also opened up the new field of discovering and characterizing new uncultivated microorganisms, which my colleagues at Stanford and elsewhere are exploring with fascinating and important results. The molecular approach we used in 1989 has subsequently revealed new medical pathogens of old diseases and also is helping to reveal the nature of uncultivated normal flora and environmental microbes that account for the vast majority of bacteria on the planet.

I mentioned that my first mentor, Stan Falkow, had changed my life. This is because 17 years after he served as my Ph.D. mentor, we married and began a partnership in research as well as in life. Our laboratory focuses on a broad range of bacterial pathogens and is dedicated to understanding the nature of the relationship between the host cell and the pathogen. Perhaps even more important to my career, the studies in my laboratory on *Helicobacter*, *Legionella*, and *Bartonella* all started as a result of the lucky chances to put the clinical, medical microbiological, and basic microbiological pieces of the puzzles together, which was afforded by this rather circuitous and unconventional career path in which I became an infectious diseases physician, a clinical microbiologist, and a scientist. I would recommend this road, even though long and sometimes arduous, to anyone who is intrigued and curious about the biology of infectious diseases. Finding the right mentor is crucial to success. For a woman training in medicine in the 1960s and 1970s, a mentor was and still is especially important. Above all, you should follow your dreams, even when you're not certain where the path will lead, because if you do, you're certain to find an interesting and rewarding career and a fulfilling life.

LUCY S. TOMPKINS was born in St. Louis, Missouri, in 1940. She attended Stanford University from 1958 to 1961 and then earned her bachelor of science degree at the University of Denver in 1963. She received her doctoral degree in 1971 at Georgetown University and her medical degree two years later at Dartmouth Medical School. She completed both an infectious diseases postdoctoral fellowship and a clinical microbiology postdoctoral fellowship at the University of Washington in 1979. After her postdoctoral work, she went on to become Assistant Professor of Microbiology and Immunology and Laboratory Medicine while simultaneously being the Associate Director and Codirector of the Clinical Microbiology Laboratory at the University of Washington until 1983. She then worked her way from Assistant Professor to Professor of Medicine (Infectious Diseases and Geographic Medicine) and Microbiology and Immunology at Stanford University School of Medicine and became the Director of the Clinical Microbiology Laboratory at Stanford University Medical Center. In 1999, she became medical liaison to that laboratory. She also is Professor of Medicine and Microbiology, Immunology, and Pathology and Medical Director of the Hospital Epidemiology and Infection Control Program at Stanford.

The following papers are representative of her publications:

Segal, E. D., J. Cha, S. Falkow, and L. S. Tompkins. 1999. Altered states: Involvement of phosphorylated CagA in the induction of host cellular growth changes by *Helicobacter pylori*. *Proc. Natl. Acad. Sci. USA* **96:**14559–14564.

Tompkins, L. S. 1992. The use of molecular methods in infectious diseases. *N. Engl. J. Med.* **327:**1290–1297.

Segal, E. D., J. Shon, and L. S. Tompkins. 1992. Characterization of *Helicobacter pylori* urease mutants. *Infect. Immun.* **60:**1883–1889.

Lowry, P. W., R. J. Blankenship, W. Gridley, N. J. Troup, and L. S. Tompkins. 1991. A cluster of *Legionella* sternal-wound infections due to postoperative topical exposure to contaminated tap water. *N. Engl. J. Med.* **324:**109–112.

Relman, D. A., J. S. Loutit, T. M. Schmidt, S. Falkow, and L. S. Tompkins. 1990. The agent of bacillary angiomatosis. An approach to the identification of uncultured pathogens. *N. Engl. J. Med.* **323:**1573–1580.

Miller, J. F., W. J. Dower, and L. S. Tompkins. 1988. High-voltage electroporation of bacteria: Genetic transformation of *Campylobacter jejuni* with plasmid DNA. *Proc. Natl. Acad. Sci. USA* **85:**856–860.

Tompkins, L. S., J. J. Plorde, and S. Falkow. 1980. Molecular analysis of R-factors from multiresistant nosocomial isolates. *J. Infect. Dis.* **141:**625–636.

MICHAEL A. PFALLER

Discovering What You Do and Don't Want

Science was not a particular area of interest for me until I enrolled in my first course, botany, at Linfield College in 1968. Before that experience my only exposure to science was high school biology and the obligatory fetal pig and frog dissection. At Linfield, I was introduced to a dynamic and charismatic individual, Dr. James Crook. Dr. Crook taught several courses in the Biology Department and did so in a manner that both excited and challenged his students. Through courses in botany, zoology, and parasitology, I became intensely interested in science and scientific investigation. As my first course in microbiology, parasitology was fascinating to me, and I have studied eukaryotic microorganisms ever since. Encountering Dr. Crook was my good fortune. He convinced me that I had an aptitude for science and microbiology and also introduced me to a couple of books, the *Microbe Hunters* and *Arrowsmith*, which fueled my interest in microbiology and infectious diseases, as they have countless other young students.

On entering medical school at Washington University in St. Louis in 1972, I immediately sought a research mentor and found one in Dr. Joseph Marr. Dr. Marr was at that time the Director of the Clinical Microbiology Laboratory at Barnes Hospital and an Associate Professor in the Division of Infectious Diseases. His research interests were in the area of metabolism of protozoan parasites, and I worked in his laboratory investigating the effects of allopurinol on the in vitro growth of *Leishmania*. This was my first exposure to drug-bug interactions and led to my first publication.

Throughout medical school, I retained an interest in infectious diseases and benefited greatly from exposure to Drs. Gerald Medoff and George Kobayashi, among others. At graduation in 1976, I was awarded the Jacques J. Bronfenbrenner Award for excellence in infectious diseases. Although I drifted away from this field for a time, I am certain that these early positive experiences strongly influenced my ultimate career choice.

After graduation from medical school, I completed an internship in surgery and worked as a general practitioner in the Indian Health Service and as an emergency medicine physician. During these three years, I found out what I did not want to do with my career and spent quite a bit of time considering what I wanted to do. The great thing about medicine as a career is that it offers so many options. I kept coming back to my positive experiences in the research laboratory and my interest in microbiology. Eventually I decided that I needed to return to the university setting and obtain additional training. Through my contacts at Washington University, I was aware of the training program in laboratory medicine. This program was unique in that it was specifically designed to train physicians for an academic career in a subspecialty area of laboratory medicine or clinical pathology. As such, it placed great emphasis on research as a component of the training program.

I decided that training in laboratory medicine was the career path that I should take and that my subspecialty area would be microbiology. At that time, I really had no concept of what clinical microbiology entailed but knew that I wanted to do what interested me—to work with "bugs."

Even before entering the Laboratory Medicine Training Program, I connected with Dr. Donald Krogstad as a possible research mentor. Dr. Krogstad had recently taken a position in the Division of Infectious Diseases and the Division of Laboratory Medicine at Washington University and along with Dr. Patrick Murray was Co-Director of the Clinical Microbiology Laboratory. Dr. Krogstad was just beginning to work with the protozoan parasite *Plasmodium falciparum*. An in vitro culture system for *P. falciparum* had just been developed by Dr. William Trager, and Dr. Krogstad was interested in refining the system and using it to study the mechanism of action and resistance to chloroquine and other antimalarial agents among strains of *P. falciparum*. This work fascinated me, so in addition to my general laboratory medicine training, I worked in Dr. Krogstad's laboratory for the next three years. We published several papers on the in vitro antimalarial activity of various agents, including a new antifungal agent, ketoconazole. This opportunity proved invaluable because it provided training in the scientific approach to problems, technical experience, and the joys and frustrations of laboratory work, publication, and grant writing.

The Laboratory Medicine Training Program proved to be exactly what I needed to focus my career objectives. Through rotations in the various clinical laboratories, I was able to clearly define clinical microbiology as my niche. During my fourth year in the program, I focused all my energies in the area of clinical microbiology under the guidance of Dr. Pat Murray. Working with Dr. Murray, I became involved in a number of "applied" research projects, including evaluations of blood culture systems, urine screening devices, and antimicrobial susceptibility testing. These projects led to eventual publications in the clinical microbiology literature and showed me how one could combine clinical service work in clinical microbiology with academic pursuits. Each aspect made the other one better. Again, throughout my training program I received great advice and guidance from Drs. Leonard Jarrett, Jay McDonald, George Kobayashi, and Gerry Medoff.

My training and experience in both clinical and laboratory medicine prepared me well for an academic career in clinical microbiology, and seven years after graduating from medical school, I finally went out and found a real job. I joined the faculty of the Department of Pathology at the University of Iowa, and except for a brief stint at the Oregon Health Sciences University, have been at Iowa ever since. My work at Iowa has evolved over the years but has remained focused on clinically relevant issues in microbiology and infectious diseases. During this time, I have continued to pursue applied research in terms of evaluation of in vitro test systems in the clinical laboratory. In addition, the drug-bug work has continued with evaluation of new antifungal agents and antibacterial agents. The topic of antimicrobial resistance became very popular in the 1990s, and our microbiology group at Iowa has been in the thick of this field of investigation. Coupled with all of these efforts has been the application of molecular tools for the study of the epidemiology of hospital-acquired infections, as well as the development of new and improved methods for both phenotypic and genotypic characterization of resistant organisms. Clearly the field of clinical microbiology is rich with interesting problems and areas of investigation. One is only limited by time and energy.

The success of any individual in academic medicine is almost always directly related to the quality of one's mentors and collaborators. At the University of Iowa, I have been extremely fortunate to be surrounded by wonderful colleagues in a supportive department and College of Medicine. Dr. Frank Koontz, clinical microbiologist extraordinaire, helped me to develop both as a microbiologist and an academician. To this day, his depth of knowledge and good humor are a source of inspiration. Likewise Drs. Ronald Jones and Gary Doern have been simply outstanding colleagues and friends. Each of these microbiologists has accomplished great things on their own yet has the wisdom to see that by working together as a team, we can all accomplish even more and have fun doing it.

Outside the Department of Pathology, I have benefited greatly by working with Drs. Richard Wenzel and Loreen Herwaldt of Hospital Epidemiology. Collaborations with these individuals and other colleagues in hospital epidemiology and infection control have been highly productive, stimulating, and fun. Again, the key has been collaboration, sharing, and friendship. The close collaboration between clinical microbiology and hospital epidemiology is one that should be sought out by all concerned. Finally, as part of an academic medical center, one always has a responsibility to train others. Our Medical Microbiology Division has been extremely fortunate to be able to provide research and clinical training opportunities to a large number of residents, fellows, and visiting scientists. Some of our most valued collaborators and friends are those individuals from other countries who elected to spend time with us in Iowa City. We provided the training opportunities and they have made the most of them. They always enrich our laboratories both in terms of culture and intellect. The opportunity to interact productively with others is the most satisfying aspect of what we do.

Clinical microbiology is a career of choice. The ways to craft a career in this field are almost limitless. From the perspective of the clinical microbiology laboratory, one can become involved in all of the most interesting infectious disease cases. Problems in diagnosis and control of infectious diseases provide a never-ending array of material for investigation and publication. Applied research in this area invariably leads to improvements in how the clinical laboratory can serve its clients and, in many cases, can serve as a springboard to more basic investigation. I am fortunate to have been involved in most of these aspects of clinical microbiology. It took me a while to choose clinical microbiology as a career, but it has been well worth it.

MICHAEL A. PFALLER received his medical degree in 1976 from Washington University School of Medicine in St. Louis, Missouri. As a medical student, he received the Bronfenbrenner Award for outstanding research in the field of infectious disease. His postdoctoral training continued at Washington University School of Medicine and included a fellowship in laboratory medicine, during which he served as Chief Resident of Laboratory Medicine, and a research fellowship in microbiology and infectious disease. In 1983, he was named Associate Director of Clinical Microbiology, University of Iowa Hospitals, and appointed Assistant Professor of Pathology at University of Iowa School of Medicine. With the exception of two years in Oregon, Dr. Pfaller's entire professional career has been at the University of Iowa, where he is now Professor of Pathology and Co-Director of Clinical Microbiology and oversees the Molecular Epidemiology and Fungus Testing Laboratory. His research accomplishments have resulted in 378 papers, seven books, and fifty-three book chapters. Although much of this work has focused on in-depth molecular epidemiology and antifungal susceptibility testing studies, his publications cover a wide array of topics significant to the field of clinical microbiology. Only his unselfishness and willingness to serve as a mentor for others equal the magnitude of his enthusiasm for clinical research.

The following papers are representative of his publications:

Jones, R. N., D. M. Johnson, M. E. Erwin, M. L. Beach, D. J. Biedenbach, and M. A. Pfaller. 1999. Comparative antimicrobial activity of gatifloxacin tested against *Streptococcus* spp. including quality control guidelines and etest method validation. Quality Control Study Group. *Diagn. Microbiol. Infect. Dis.* **34:**91–98.

Pfaller, M. A., R. N. Jones, G. V. Doern, H. S. Sader, K. C. Kugler, and M. L. Beach. 1999. Survey of bloodstream infections attributable to gram-positive cocci: Frequency of occurrence and antimicrobial susceptibility of isolates collected in 1997 in the United States, Canada, and Latin America from the SENTRY Antimicrobial Surveillance Program. *Diagn. Microbiol. Infect. Dis.* **33:**283–297.

Seguin, J. C., R. D. Walker, J. P. Caron, W. E. Kloos, C. G. George, R. J. Hollis, R. N. Jones, and M. A. Pfaller. 1999. Methicillin-resistant

Staphylococcus aureus outbreak in a veterinary teaching hospital: Potential human-to-animal transmission. *J. Clin. Microbiol.* **37:**1459–1463.

Pfaller, M. A., J. Zhang, S. A. Messer, M. E. Brandt, R. A. Hajjeh, C. J. Jessup, M. Tumberland, E. K. Mbidde, and M. A. Ghannoum. 1999. *In vitro* activities of voriconazole, fluconazole, and itraconazole against 566 clinical isolates of *Cryptococcus neoformans* from the United States and Africa. *Antimicrob. Agents Chemother.* **43:**169–171.

Klepser, M. E., R. E. Lewis, and M. A. Pfaller. 1998. Therapy of *Candida* infections: Susceptibility testing, resistance, and therapeutic options. *Ann. Pharmacother.* **32:**1353–1361.

Wendt, C., S. A. Messer, R. J. Hollis, M. A. Pfaller, and L. A. Herwaldt. 1998. Epidemiology of polyclonal gram-negative bacteremia. *Diagn. Microbiol. Infect. Dis.* **32:**9–13.

Pfaller, M. A., S. R. Lockhart, C. Pujol, J. A. Swails-Wenger, S. A. Messer, M. B. Edmond, R. N. Jones, R. P. Wenze, and D. R. Soll. 1998. Hospital specificity, region specificity, and fluconazole resistance of *Candida albicans* bloodstream isolates. *J. Clin. Microbiol.* **36:**1518–1529.

Jones, R. N., and M. A. Pfaller. 1998. Bacterial resistance: A worldwide problem. *Diagn. Microbiol. Infect. Dis.* **31:**379–388.

Wasilauskas, B., R. Gay, P. Zwadyk, M. Pfaller, and F. Koontz. 1987. Multicenter comparison of MicroScan and BACTEC blood culture systems. *J. Clin. Microbiol.* **25:**2355–2358.

Pfaller, M., B. Ringenberg, L. Rames, J. Hegeman, and F. Koontz. 1987. The usefulness of screening tests for pyuria in combination with culture in the diagnosis of urinary tract infection. *Diagn. Microbiol. Infect. Dis.* **6:**207–215.

Did I Choose Clinical Microbiology or Did It Choose Me?

The United States was already six months into World War II when I graduated from Colorado College in June 1942. I had little opportunity to think about or plan a career in anything. A persuasive correspondence from President Roosevelt ordering me to report for military service eliminated the need to choose. My career would have to wait. After induction into the U.S. Army and basic training, I was sent to Fitzsimons General Hospital for training in medical laboratory technology. This was a lucky break because it allowed me to use some of what I had learned as an undergraduate biology major. I gravitated toward the diagnostic microbiology segment of laboratory technology, and after four months of training I was assigned to Patton's 3rd Army with duty in the microbiology laboratories of field and general hospitals in the European theater.

On discharge from the military, I enrolled in the graduate school of Syracuse University because I was interested in fermentation and industrial microbiology, which were major interests of the microbiology department there. My interest stemmed from my observations of the use of penicillin in the miraculously successful treatment of serious bacterial infections in the military and civilian population in France, Germany, and elsewhere. I thought a career in fermentation microbiology would be interesting and productive. This proved not to be the case, and after satisfying the requirements for a master's degree, I transferred to the University of Kentucky, which had a well-known and popular microbiology department that emphasized medical microbiology.

This was a wise move, and after receiving a Ph.D., I joined the staff of the Lexington Clinic (40 physicians practicing group medicine) to oversee their laboratory and set up a productive and active infectious disease laboratory. Postdoctoral training programs for microbiologists were virtually nonexistent. To get the training and exposure I needed to do my job, I managed a "visiting fellowship" to the clinical laboratories of the Mayo Clinic in Rochester, Minnesota, and another "fellowship" to the Cleveland Clinic to gain experience in the laboratory aspects of blood and blood products transfusions. These fellowships proved to be valuable experiences for me, and I was able to establish a first-class microbiology laboratory and blood bank at the Lexington Clinic and St. Joseph Hospital in Lexington, Kentucky.

Shortly after, in 1960, the University of Kentucky Medical Center opened and the faculty and staff were in place. I was fortunate to be appointed associate professor in the Department of Medicine. My job was to teach medical students, to provide diagnostic service in infectious diseases for the clinicians, and to engage in research. At the same time I maintained my positions at the Lexington Clinic and St. Joseph Hospital. This was a worthwhile "town and gown" arrangement, and although I was quite busy, I managed to successfully accomplish what was expected.

My research interests were broad rather than focused on one disease or one organism or group of organisms. We showed that group B β-hemolytic streptococci were capable of producing serious systemic disease in children and postpartum women. Previously, group B streptococci were not seriously considered to be pathogens; many such infections were incorrectly diagnosed as being caused by group A streptococci. Initially, this finding did not alter therapy, but over time, antibiotic resistance patterns in group B were found. Also, the epidemiology of group A and group B streptococcal infections is different.

I was also interested in systemic fungal diseases that were peculiar to the geographical area. Two such diseases were histoplasmosis and blastomycosis—both endemic to the Missouri Valley. We confirmed the presence of *Histoplasma capsulatum* in hen house droppings and soil. We showed that blastomycosis occurs frequently in dogs that may have transmitted the infection to humans by contact, but we were unable to prove conclusively that soil is the natural habitat of *Blastomyces dermatitidis*. We were among the first to show the great genetic diversity among clinical isolates of *Escherichia coli* strains with unusual morphological, biochemical, or antigenic characteristics. These were sent to the Centers for Disease Control and Prevention (CDC) for confirmation; these strains, together with those that the CDC received from other laboratories, provided the groundswell that resulted in showing that certain *E. coli* strains with or without aberrant phenotypic characteristics produce endotoxins that cause diarrhea or systemic infections involving other organs.

At this same time—the early 1960s—a small but growing number of microbiologists who worked in hospital or medical clinics, clinical laboratories, or in state public health laboratories began to ask questions or raise issues within the American Society for Microbiology (ASM) organizational structure that dealt with the problem of professional recognition. This group within ASM introduced the designations of *clinical microbiology* and *clinical microbiologists*. Clinical microbiologists were described as microbiologists who were in charge of or worked in a clinical laboratory that received and processed clinical specimens to establish a laboratory diagnosis of an infectious disease and determine a profile of antibiotics to which the isolated etiologic agent would respond.

At this time, there were approximately eight or ten marketed antibiotics and about four sulfadiazine derivatives that were approved for treating various bacterial infections. It became increasingly important to isolate, identify, and determine the antibiotic susceptibility of the pathogen(s) as quickly and accurately as possible. Unfortunately, the means of communicating and exchanging information and the use of "standardized" methods were either nonexistent or very limited. To establish communication within the ASM framework, I organized a round table, "Current Trends in Clinical Microbiology," for presentation at the annual general meeting. This round table was approved by the Program Committee and became part of the 1964 ASM annual meeting program. Henry Isenberg subsequently joined me as coconvener, and together we managed to organize seminars that addressed timely topics in clinical microbiology. These were extremely popular because they provided an open forum for exchange of information and discussions that ultimately led the way to considering ways and means of disseminating this information in a publishable form.

I took the initiative and started the *Clinical Microbiology Newsletter*, which originated from my home and office and was funded by Baltimore Biological Laboratories, as the ASM had no interest in this activity. Similarly, ASM was not interested in publishing any of the seminar proceedings. I negotiated a contract with Charles C. Thomas, Publisher, who agreed to publish and offer for sale an expanded version of the topics presented at the clinical microbiology seminars. About 20 books on a wide variety of topics were published over the next 20 years; they became popular and extensively used in clinical microbiology laboratories.

Finally, after much debate, argument, and persuasion, the ASM recognized clinical microbiology as a microbiologic discipline and the Clinical Microbiology Division was approved. We were able to convince the ASM hierarchy to undertake publishing a *Manual of Clinical Microbiology* (MCM). ASM agreed to this with tongue in cheek, believing that the book would fail financially and not be accepted or purchased by clinical microbiologists. MCM*1* was an overwhelming success; it required three printings to meet the demand.

The ASM Council also approved the publication of the *Cumitech* series with John Sherris as the founding editor, and last, but not least, the council also approved the establishment of a section in the journal *Applied Microbiology* to be devoted to clinical microbiology papers that met the editors' criteria. I was appointed editor of this section; it was first published in January 1970. The success of this section exceeded all expectations, and eventually it was split off and published as the *Journal of Clinical Microbiology* (JCM). The name of *Applied Microbiology* was changed to *Applied and Environmental Microbiology*. I was appointed editor-in-chief of JCM, and after the initial planning, organizing, and appointments of the editors and editorial review board, JCM was first published in January 1975. JCM is now in its twenty-fifth year, the seventh edition of MCM was published in 1999, and the Cumi-Tech series continues to break ground in developing and publishing at least 31 technical manuals on timely topics in clinical microbiology. These have all been scientific and technical successes and have generated extraordinary profits for ASM.

At the same time these publishing ventures were undertaken, several of us in the Clinical Microbiology Division organized and presented workshops just before the general meeting. The workshops were also profitable and the demand for them increased; it became necessary to plan for an increased number of workshops on timely topics. The Board of Education and Training later became responsible for the pre- and post-annual meeting workshops. They are now an integral part of the program and cover subjects beyond clinical microbiology—for example, environmental issues, biosafety, and molecular microbiology.

At this time (circa 1960–1970), the number of antibiotics that were discovered or synthesized increased logarithmically, which created a pressing need for standardized rapid diagnostic techniques and in vitro antibiotic susceptibility testing. These needs stimulated a beehive of activity in the biomedical industry, pharmaceutical companies, and many clinical microbiology laboratories and led to collaborative research between industry and clinical microbiology laboratories on new reagents and instrumentation technology. To achieve standardization, I contacted a group of colleagues to initiate a program of interlaboratory testing of methods and instruments. We were successful in our efforts, presented our findings at ASM meetings, and published them in acceptable fashion.

The evolution of standardized methods led to the recognition of the need for quality control—a concept that generally was not practiced—in clinical microbiology laboratories. Again, collaboration between industry and clinical microbiology laboratories was necessary. I was able to organize several round tables and symposia that addressed quality-control measures and practices and the necessary reagents, equipment, and published procedures that promoted quality control and quality assurance in clinical laboratories. As a result, laboratory performance improved considerably, and laboratory data were increasingly meaningful and useful to clinicians and surgeons in patient care.

During the early 1960s, the National Committee for Clinical Laboratory Standards (NCCLS) became prominent. It was a direct outgrowth of the collective recognition for standardization in clinical laboratories. I was appointed chair of the NCCLS Microbiology Subcommittee with a clear mandate to investigate and establish standardized in vitro susceptibility testing techniques. Kirby, Bauer, Sherris, et al. had made considerable progress in the standardization of an in vitro drug-impregnated disc method for testing antibiotic susceptibility. This was a well-developed, tested, and documented procedure, but it required an elaborate multilaboratory testing and evaluating effort to make their procedure (or an acceptable modification) acceptable as an NCCLS standard. I recognized this as a high priority for the NCCLS

Microbiology Subcommittee. It proved to be much easier said than done, but the task was undertaken and ultimately completed. The first NCCLS standard on in vitro antibiotic susceptibility testing was published in 1975. I presented and distributed 500 copies of this standard at an international congress in London that same year.

In January 1969, I was presented with the opportunity to become the Director of the Bacteriology Division of the Communicable Disease Center (later to become the CDC). In July 1969, I moved my family to the greater Atlanta area and so ended the first half of my professional career as a clinical microbiologist. On September 1, 1969, I reported to the Bureau of Laboratories, CDC, and began the second part of an exciting and challenging career in the U.S. Public Health Service. Nevertheless, that's another story for another day.

ALBERT BALOWS was born in Denver, Colorado, in 1921, but grew up in Colorado Springs. He received his bachelor's degree from Colorado College and his master of science degree from Syracuse University in 1948. He received his doctoral degree in microbiology from the University of Kentucky in 1952. He served in the U.S. Army from 1942 until 1946. His military experience included a tour of duty in a series of military hospitals in France, Germany, and Belgium and finally in Paris, as well as medical laboratory training at Fitzsimons General Hospital and the Institut Pasteur. After his military service, he received postdoctoral training at the Mayo Clinic in Rochester, Minnesota, the Cleveland Clinic, and the Ortho Research and Education Foundation and additional training in environmental microbiology at the National Centers for Disease Control and Prevention (CDC). Dr. Balows' academic appointment credentials are impressive. From his first appointment as an Instructor in the Department of Plant Sciences (Microbiology) at Syracuse University through his most recent position as a Clinical Professor of Pathology and Laboratory Medicine at Emory University School of Medicine, he has contributed enormously to the understanding of microbiology and infectious diseases. The laboratory phase of his clinical microbiology career began in 1950, when he assumed directorship of the microbiology and blood bank departments at St. Joseph's Hospital in Lexington, Kentucky. In 1969, he left Kentucky and embarked on his long and distinguished career at the CDC in Atlanta, where he was Chief of the Bacteriology Section and Branch from 1969 to 1974, Director of the Bacteriology Division from 1974 to 1981, director of the Sexually Transmitted Diseases Laboratory Program from 1981 to 1983, and Assistant Director for Laboratory Science from 1981 to 1988. He was Adjunct Professor at Emory University School of Medicine and Georgia State University. He is a Past President and an honorary member of the American Society for Microbiology (ASM). He was Chair of the Board of Governors of the American Academy of Microbiology. His research interests have been in applied and public health microbiology in the rapid diagnosis, therapy, and control of infectious diseases. He has published over 95 book chapters and extensively in peer-reviewed journals. Dr. Balows was the founding Editor-in-Chief of the *Journal of Clinical Microbiology* and served on 12 journal editorial boards. In addition, he was the Editor for the third and fourth editions of the *Manual of Clinical Microbiology* and the Editor-in-Chief for the fifth edition. He was also the Editor-in-Chief for *The Prokaryotes* and an Editor for Topley and Wilson's ninth edition of *Microbiology and Microbial Diseases*. He is the first Past President of ASM to receive the American Board of Medical Microbiology Professional Recognition Award, the Becton Dickinson and Company Award in Clinical Microbiology, The Abbot Laboratory Outstanding Achievement Award, and countless other awards for his contributions to and leadership in clinical microbiology. He was the recipient of the 1999 bioMérieux Sonnenwirth Award in recognition of his exemplary leadership in clinical microbiology. Specifically, he was honored for his promotion of innovation in clinical laboratory science, for high dedication and commitment to ASM, and for advancement of clinical microbiology as a profession. Dr. Balows has served as Chair of the American Academy of Microbiology. He likewise chaired the Clinical Microbiology Division of ASM.

The following books and papers are representative of his publications:

Balows, A., W. J. Hausler, M. Ohashi, and A. Turano (ed.). 1988. *Laboratory Diagnosis of Infectious Diseases: Principles and Practice: Bacterial, Mycotic, and Parasitic Diseases*. Springer-Verlag, New York, N.Y.

Reeves, M. W., L. Pine, J. B. Neilands, and A. Balows. 1983. Absence of siderophore activity in *Legionella* species grown in iron-deficient media. *J. Bacteriol.* **154:**324–329.

Blaser, M. J., F. W. Hickman, J. J. Farmer 3d, D. J. Brenner, A. Balows, and R. A. Feldman. 1980. *Salmonella typhi*: The laboratory as a reservoir of infection. *J. Infect. Dis.* **142:**934–938.

Smith, P. B., T. L. Gavan, H. D. Isenberg, A. Sonnenwirth, W. I. Taylor, J. A. Washington 2d, and A. Balows. 1978. Multi-laboratory evaluation of an automated microbial detection/identification system. *J. Clin. Microbiol.* **8:**657–666.

Balows, A. 1978. Training and certification of clinical microbiologists. *Ann. Internal Med.* **89:**812–814.

Balows, A. 1977. An overview of recent experiences with plasmid-mediated antibiotic resistance or induced virulence in bacterial diseases. *J. Antimicrob. Chemother.* **3:**3–6.

Brenner, D. J., and A. Balows. 1975. Evaluation of the Enteric Analyzer, an instrument to aid in the identification of Enterobacteriaceae. *J. Clin. Microbiol.* **2:**235–242.

Balows, A. (ed.). 1974. *Anaerobic Bacteria: Role in Disease.* Charles C Thomas, Springfield, Ill.

Starr, S. E., F. S. Thompson, V. R. Dowell, Jr., and A. Balows. 1973. Micromethod system for identification of anaerobic bacteria. *Appl. Microbiol.* **25:**713–717.

JOSEPHINE A. MORELLO

Evolution of a Clinical Microbiologist

The inner-city environment of the West End of Boston, where I grew up, was an unlikely stimulus for an interest in the natural sciences. Although I was headed for the then traditional woman's career of elementary school teacher, a high school biology course fascinated me as I learned about the interrelationships and intricacies of biological systems. This interest was narrowed to the field of microbiology during my undergraduate years as a biology major at Simmons College; it was great fun and a challenge to work with microorganisms from a variety of sources and especially to learn about those agents involved in disease production. While attending graduate school at Boston University's Division of Medical Sciences, with the intent of completing a master's degree only, I was astonished when I was encouraged to apply to the doctoral program, was accepted, and received a fellowship to complete my studies. Overcoming my qualms about surviving the rigors of doctoral studies, I proceeded with my course work and research.

The Microbiology Department in which I studied was small and somewhat "low tech"; graduate students worked independently, performing all tasks required for their work, including washing dishes and making media. How quickly I performed these tasks often dictated how many experiments I could perform each week. My doctoral research involved studying interactions of *Salmonella* with phagocytes in vitro. The aim of this work was to study the cellular immune response to salmonellae, first by using mouse cells and then moving on to a human system (which I never did). At that time, knowledge of the intricacies and complexities of the immune system was rudimentary, and my work did not clarify the inconsistencies already present in the literature regarding the intracellular fate of microorganisms. I note, however, that although significant advances in that area have continued to be made, many of the same questions with which I struggled remain unanswered.

After completing my degree, I remained at Boston University for two years to study chronic salmonellosis in mice, a project funded by the National Cancer Institute. Long-term drug studies were hampered by deaths in mice with occult *Salmonella* infections, and a way was needed to detect infection before trials were begun. A master's degree student and I worked together and developed an accurate serologic assay for this purpose. More significantly, my student was also the supervisor of the University Hospital microbiology laboratory, and during our work, which also involved culture detection of the enteric infection, she stimulated my interest in the diagnostic aspects of microbiology.

Soon I decided it was time to leave Boston and explore other opportunities. During my investigative work, I began to suspect that a career devoted only to research would not prove sufficiently satisfying for me, and I wondered about my choice of profession. Because of my new interest in diagnostic microbiology, I contemplated applying to the clinical microbiology postdoctoral training programs that were now being offered. Instead, the opportunity arose to take a position as a research associate at the world-famous Rockefeller Institute (now Rockefeller University) in New York City, an opportunity that seemed too significant to turn down. Here my project was on the nature of the antigenic stimulus, particularly immunologic properties of phagocytized antigen and its role in the anamnestic response. Working at the "high-tech" institute was an amazing change from my previous research experience. The intensity, drive, and commitment of the researchers were remarkable, and I came to know several young people who have since made outstanding scientific discoveries. However, my frustration with the slow speed of such discoveries again made me contemplate my career direction. Instead of continuing on there, I left Rockefeller Institute after two years to pursue a clinical microbiology postdoctoral program at Columbia University College of Physicians and Surgeons with Dr. Paul D. Ellner. This decision was made in spite of the low opinion of the field at this time, an opinion that prompted one of my most important mentors in Boston to write that I was "throwing away my doctoral degree and becoming an overqualified technician." (A number of years later, I received a sincere apology.)

No matter. At last I knew what I wanted to be when I "grew up." Something this much fun that gave me such an immense sense of accomplishment couldn't be as bad as reputed! Every day the laboratory provided critical clinical information that affected medical decisions and was of direct patient benefit. The medically oriented courses I had taken during graduate school provided an excellent foundation for my training. In addition to learning basic laboratory procedures, attending infectious disease conferences, and taking some courses to supplement my experience, I worked on the development of Columbia broth, an enriched blood culture medium that came into widespread clinical use for a time. Both at Boston University and Columbia University, I also had the opportunity to participate in microbiology courses for medical students, a teaching experience that satisfied my very first career ambition, albeit at a different level.

After completing this postdoctoral training, I became Director of Microbiology at Harlem Hospital Center, one of the City Hospitals of New York, which recently had become affiliated with Columbia-Presbyterian Hospital. This exciting and sometimes daunting challenge provided invaluable experience not only in infectious disease diagnosis but also in other matters that, during the past thirty years, have become increasingly prominent in the medical field: administrative and fiscal responsibility (dealing with the bureaucratic City of New York!) and cultural diversity. In spite of the unique opportunity afforded by this position, within two years I received an "offer that I couldn't refuse" from the University of Chicago and headed to Chicago. There I not only entered a new phase of my professional life but also of my personal life, for soon I met the person who would become my husband and who has provided indispensable support for my diverse activities.

In Chicago, I finally found a use for all my education and training and was fully accepted into the academic community. I had the opportunity to develop a state-of-the-art clinical microbiology laboratory, teach undergraduates and medical students, train other doctoral-level clinical microbiologists, and undertake clinical and basic science research programs. As had been evident from the negative comments I received when entering the field, the discipline of clinical microbiology was not highly regarded, and unlike other areas of laboratory medicine, was not well standardized. Early on, several clinical microbiologists began to assess critically microbiological procedures that either were already in place or coming into use. They also championed the publication of journals and manuals specifically devoted to this discipline. These efforts were aimed at providing a rational basis for clinical microbiology laboratory work and to gain respect for the field.

My first few years at Chicago paralleled the early development of rapid methods and automation for clinical microbiology, and I was able to work with several systems and validate their use in the clinical laboratory. The areas in which I was especially interested were blood culture instruments and automated and nonautomated rapid identification systems, especially for blood culture isolates and anaerobes. These types of studies were important for determining which systems were accurate, cost-effective, and easily integrated into the work flow of the laboratory. They also provided information for smaller laboratories that were not able to conduct such extensive studies.

Shortly after arriving at the University of Chicago, I met and began working with Marjorie Bohnhoff, a research associate whose previous collaborator had retired. We embarked on a study of the characteristics of strains of *Neisseria gonorrhoeae*. At that time (the early 1970s), the gonorrhea epidemic was in full swing, and eventually we collected approximately 1,200 strains from our patient population. Among these we selected about 250 strains for study, half of which had been collected from patients with disseminated gonorrhea and the other half from patients with local infection only. Using several methods (auxotyping, immunologic characteristics, and susceptibility to penicillin), we found distinct differences in these parameters among strains from the two types of infections. These included a disproportionate nutritional requirement for arginine, hypoxanthine, and uracil (AHU^-) among strains from disseminated infections, which was accompanied by increased susceptibility to penicillin. These studies helped differentiate this unique population and permitted investigations of additional factors related to the pathogenesis of these strains.

Along with Dr. Stephen Lerner, an infectious disease specialist, and Dr. William Janda, a clinical microbiology postdoctoral fellow, we conducted other studies characterizing strains from disseminated and localized infection, determined a unique biochemical pathway for the AHU$^-$ strains, and studied gonococcal and meningococcal strains isolated from homosexual men. It was exciting to be working in the research laboratory again while remaining actively involved in the clinical laboratory.

For a number of years, I was able to secure funding to train several postdoctoral fellows in clinical microbiology. These included Drs. Paul Stiffler, Raymond Kaplan, William Janda, Margaret Bartelt, Mary York, and Shan-Ching Ying. Each has gone on to make his or her own contributions to the field of clinical microbiology, and Dr. Ying, a pathologist, is now responsible for ensuring high-quality microbiology testing in a community hospital setting.

In addition to academic and clinical laboratory responsibilities, another important facet of my career has been participation as a volunteer in the activities of several professional societies, particularly the American Society for Microbiology (ASM). Serving on committees and participating in workshops and seminars provided a singular opportunity to interact with other clinical microbiologists and colleagues from diverse disciplines of microbiology, many of whom have become long-standing friends. Perhaps my most rewarding, though intensive, volunteer experience was as the first editor-in-chief of ASM's then newest journal, *Clinical Microbiology Reviews*. This ten-year experience was documented in an article in *Clinical Microbiology Reviews* prepared in conjunction with ASM's centennial activities.

In my thirty years as a clinical microbiologist, I have seen the field grow, develop, and gain respectability. Now, however, the many changes in health care are mandating a drastic reappraisal of how to continue high-quality laboratory work with limited resources. Many people who work in health care are discontent, and only those who view the new constraints as opportunities and/or challenges to extend their expertise in new directions may succeed. Clearly, the field of clinical microbiology is undergoing changes that could not have been foreseen when I first entered it. I expect that like the forerunners who were active in the early days of my career, the new generation of clinical microbiologists will find ways to meet the challenges and bring further regard and advancement to this discipline.

JOSEPHINE A. MORELLO was born in Boston in 1936. She majored in biology at Simmons College and received a doctoral degree in microbiology in 1962 from Boston University. She was a Research Associate at the Rockefeller Institute in New York and did postdoctoral training in clinical microbiology at Columbia University College of Physicians and Surgeons in New York from 1966 to 1968. She then joined the faculty of Columbia University as Assistant Professor of Microbiology in 1968. This was a stepping stone to becoming the Director of Microbiology at the Harlem Hospital Center during the same time. She moved to the University of Chicago in 1970, working her way from Assistant Professor of Pathology and Medicine to her present positions of Professor, Director of Clinical Microbiology Laboratories, Director of Hospital Laboratories, and Vice Chair of the Department of Pathology. She has taught microbiology to undergraduates and approximately 2,000 medical students and has trained six postdoctoral fellows in clinical microbiology. She has also done laboratory and administrative work as a clinical microbiologist and researched the characteristics of *Neisseria gonorrhoeae* and rapid methods and instrumentation for the clinical laboratory. She has received the Alexander Sonnenwirth Memorial Lecturer Award and the Distinguished Service Award from the American Society for Microbiology, the Professional Recognition Award from the American Board of Medical Microbiology, and the Pasteur Award and Tanner-Shaughnessy Award from the Illinois Society for Microbiology.

The following papers are representative of her publications:

Morello, J. A. 1999. *Clinical Microbiology Reviews*: Genesis of a journal. *Clin. Microbiol. Rev.* **12:**183–186.

Morello, J. A., C. Leitch, S. Nitz, J. W. Dyke, M. Andruszewski, G. Maier, W. Landau, and M. A. Beard. 1994. Bacteremia detection by the ESP blood culture system. *J. Clin. Microbiol.* **32:**811–818.

Morello, J. A., S. M. Matushek, W. M. Dunne, and D. B. Hinds. 1991. Performance of a BACTEC nonradiometric medium for pediatric blood cultures. *J. Clin. Microbiol.* **29:**359–362.

Sahm, D., S. Boonlayangoor, and J. A. Morello. 1987. Direct susceptibility testing of blood culture isolates with the AutoMicrobic System (AMS). *Diagn. Microbiol. Infect. Dis.* **8:**1–11.

Janda, W. M., J. A. Morello, S. A. Lerner, and M. Bohnhoff. 1983. Characteristics of pathogenic *Neisseria* spp. isolated from homosexual men. *J. Clin. Microbiol.* **17:**85–91.

Morello, J. A., S. A. Lerner, and M. Bohnhoff. 1976. Characteristics of atypical *Neisseria gonorrhoeae* from disseminated and localized infections. *Infect. Immun.* **13:**1510–1516.

Morello, J. A., and P. D. Ellner. 1969. New medium for blood cultures. *Appl. Microbiol.* **17:**68–70.

Morello, J. A., and E. E. Baker. 1965. Interaction of *Salmonella* with phagocytes *in vitro*. *J. Infect. Dis.* **115:**131–141.

Morello, J. A., T. A. DiGenio, and E. E. Baker. 1964. Evaluation of serological and cultural methods for the diagnosis of chronic salmonellosis in mice. *J. Bacteriol.* **88:**1277–1282.

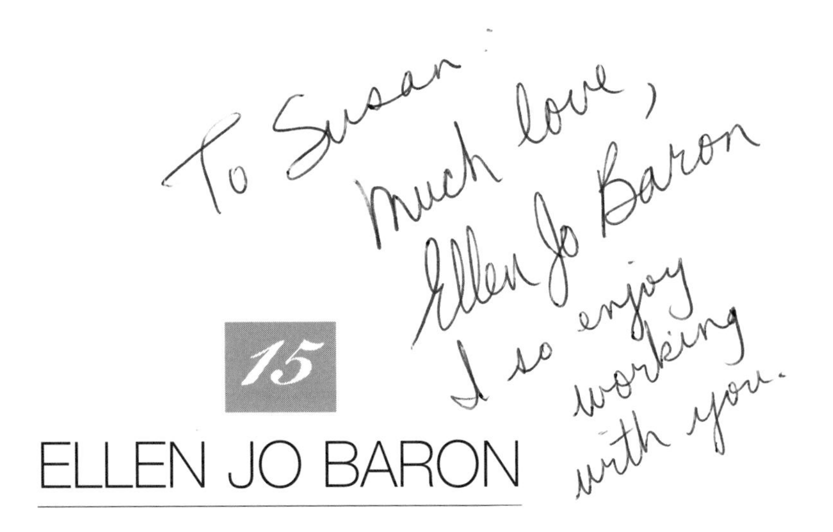

15

ELLEN JO BARON

Maturing into a Diagnostic Microbiologist for Today

Although I originally had aspirations to become an elephant trainer, that career path became less appealing once I passed age eight. Becoming a great watercolor or pen and ink artist, however, seemed reasonable. My early artistic talent was nurtured in a special program for young artists at the Carnegie Institute of Technology in Pittsburgh, Pennsylvania, on Saturday mornings during the school year in the auditorium of the Carnegie Mellon Museum of Natural History. Once my lessons were finished, I amused myself for hours wandering among the dinosaur skeletons, Egyptian mummies, and stuffed mammals. My interest in things scientific must have been born there.

A world-changing event occurred in October 1957 when I was in the fifth grade. Russia launched the first globe-circling satellite, *Sputnik*. This tiny sphere of nailed-together metal bits also launched the careers of many of my fellow scientists, as national pride and fear of defeat in the Cold War propelled many bright students to change career goals and study "science." I changed my focus from art and became the first student of science in my family's history.

After my family moved to Mitchell, South Dakota, for my father's business, I was blessed with an outstanding high school biology teacher,

Dr. William Houk. I worked every summer as a pathologist's assistant at a local hospital, washing used tubes of blood and doing frog pregnancy tests. Being from South Dakota had its advantages. As one of only two National Merit Scholars in the state the year I graduated from high school, many universities and colleges recruited me. I chose Michigan State University in East Lansing because of the large diversity of class topics and majors available; the presence of a huge student population in which I could be anonymous, as I couldn't be in my home town of tiny Mitchell; and the exposure to the wider diversity of people.

At Michigan State I majored in preveterinary science, premedical science, biology, art, and maybe something else. When it was time to graduate, I still had not chosen my major. My time and money were about to run out. I had to find the major that required the fewest additional classes. Medical microbiology won—a pragmatic choice that proved one of the most crucial of my life. Within microbiology, I discovered the visual aesthetics of microscopy and the ability to describe and identify bacteria. I could satisfy both my scientific and artistic sides. And fortunately, contrary to my classmates, I found the smells of autoclaved media and growing bacteria exotic and appealing.

When I graduated, I took a job outside the field of microbiology doing biochemical tests on animal tissues. It didn't last long. Sexual harassment by my boss forced me to search for different employment and propelled me back into microbiology. A diagnostic microbiology laboratory at a county public health department had an opening for a person with a B.S. in microbiology; I smelled the autoclave and the unique odors of a busy microbiology laboratory and knew I was home. I rotated through many areas of the laboratory, learning the basics from the ground up—*Mycobacterium tuberculosis* laboratory, urines, susceptibilities, wounds, genital cultures, and others.

When the opportunity arose, I moved into a hospital clinical laboratory in Madison, Wisconsin. There I learned all aspects of clinical microbiology from my coworkers, who taught me the importance of the patient specimens entrusted to us. The hospital sent me to the Centers for Disease Control and Prevention (CDC) to take a two-week course in hospital epidemiology, a fledgling science in those days. I was able to help start one of the first infection control programs in Madison. Working with nurses, physicians, and administrators was a great experience, one that began to shape my appreciation for multidisciplinary teams and group processes.

But I was still at the bottom rung of the clinical profession. An incident involving an anesthesiologist led me to seek a higher academic degree—one that I hoped would earn me respect and influence. This anesthesiologist controlled the postappendectomy care for a young patient and demanded antibiotic susceptibility tests on all organisms recovered from the patient's respiratory secretions. I knew that the organisms we recovered were simply the patient's normal mucosal and saliva flora and that antibiotic therapy to try to remove them was inappropriate. Nevertheless, the anesthesiologist would not listen to a lowly laboratory technologist, and the pathologists didn't have the microbiological background to refuse the doctor's requests. Multiple successive antibiotics were prescribed to treat that patient's "pathogens," and as a result the patient developed pneumonia with a panresistant *Pseudomonas aeruginosa* and died. I was devastated. I thought that if I had a master of science degree I could gain credibility and respect from clinicians. So I entered the master's program at the University of Wisconsin with the hospital's blessing, taking classes during the day and working nights. I published my first two papers on a case observation and a project carried out at the hospital, which whetted my appetite for seeing my observations in print.

After two years, I received a master's degree in medical microbiology and looked forward to the enhanced credibility with physicians that this academic credential would provide. There was none. I soon realized that only another doctor could talk to doctors on their own level. Therefore, I applied to the University of Wisconsin graduate school once again, this time for entry into the doctoral program. However, here again, being a woman was a hurdle. In my interview with the chair of the department, I was told that "as long as there are men with families to support who want to enter our program, I will not allow you to take up space in my classrooms." I was bitterly disappointed and suddenly disillusioned with my career choice.

I accepted a job at the Wisconsin State Laboratory of Hygiene, one of the premier microbiology laboratories in the United States, boasting a Ph.D. director who urged me to become an American Academy of Microbiology–registered microbiologist. Women were moving forward, especially in the clinical microbiology field. I studied diligently and passed the registered microbiologist examination, became a member of the American Society for Microbiology (ASM), and began to appreciate the community of microbiologists—one that strove to allow women like myself to advance. At the State Laboratory, regular bench workers had the opportunity to learn new things and to initiate new methods; we also learned how the literature can help shape practice. After researching the studies published on female vaginal microbiota, I finally appreciated the value of evidence-based microbiology and changed my tradition-bound way of thinking about diagnostic microbiology.

A colleague and I presented a paper at the 1976 ASM annual meeting in Atlantic City. I was exposed to the exhibit hall and the myriad presentations. It was exciting and motivating but also depressing because I didn't know anyone and spent evenings alone in my room eating potato chips and olives. Inspired by the talks I had heard at the ASM meeting, I volunteered to give a lecture at the Wisconsin State Laboratory. My first experience speaking to a group of colleagues was rewarding. I wanted to do more—I wanted my doctorate.

But again, because I was a woman it was difficult to enter the Ph.D. program at Wisconsin. Even though the chair had changed and women were now more acceptable, men still received preferential treatment. Although some male colleagues could keep their jobs and attend school part-time, I was told I could not pursue an advanced degree on a part-time basis and keep my job at the State Laboratory. I had to quit my job to go to school full-time. My male counterparts did not suffer this hardship. I made my way through graduate school on limited resources. I washed, sterilized, and reused the same box of yellow plastic pipette tips through my three years of research! I carry that spirit of frugality into my present position, a valuable attribute in the current times of constantly diminishing reimbursement for health care costs. While in retrospect, the ability to focus full-time on study and research and to learn the value of frugality was beneficial, at the time I deeply felt the sting of gender discrimination. As a woman I had to struggle more to reach my goals.

My graduate studies progressed along multiple lines. I gained valuable skills and experience as a teaching assistant that were instrumental in my career development, and when one of the professors died suddenly, I took over responsibility for the class and have been involved in teaching ever since. My major professor, Richard Proctor, an infectious disease physician of great scientific acumen, envisioned and promoted a research career for me. He also introduced me to the world of infectious disease physicians in clinical practice. I saw firsthand the contribution a hospital microbiology laboratory made to their work. That insight and an inspiring seminar by a visiting scholar, Ray Kaplan, rekindled my love for diagnostic microbiology. I chose to do a training program at the UCLA-Wadsworth Veterans Administration Medical Center that was approved by the American Academy of Microbiology and headed back toward a clinical laboratory career. My selection of the UCLA-Wadsworth program was based on several factors, including the reputation of Sydney Finegold (who led the program), the ability to participate in infectious diseases patient rounds every day, and the presence of relatives in Los Angeles. An unexpected benefit of my choice of programs was meeting my future husband, Jim Taylor. As a consultant and author in the field of organizational behavior, Jim knew a great deal about management and supervision, and our long discussions have helped shape my ideas about running a laboratory.

Nearing completion of my postdoctoral fellowship, I was worried about the lack of jobs advertised. I began contacting some of the leaders in diagnostic microbiology. Within days, Henry Isenberg called to inform me of a job opening in the New York City area. North Shore University Hospital on Long Island hired me to direct a 27-person laboratory. My time in New York helped to build my education in life as well as in microbiology. My pathologist colleagues were wonderful partners who basically left me alone to run the laboratory, though they supported my decisions. My laboratory manager, Margie Berman, taught me many lessons about proper behavior in New York and about managing people, including the importance of never saying never and not giving bad news on Friday—tenets I still follow religiously.

Although I had left Los Angeles, my ties to that city did not evaporate. Thus, after five wonderful years in New York City, and with much prodding from Jim, Sid Finegold successfully recruited me back to Los Angeles to direct his clinical research laboratory at the UCLA-Wadsworth VA Medical Center.

Taxonomy became a real interest. We recognized and characterized a new organism, *Bilophila wadsworthia*, and described the etiologies of various phases of appendicitis beyond all previous studies. Because of my computer literacy, I was positioned to develop a database for the Wadsworth anaerobe collection. Computer competence has been a major keystone of my career productivity, especially during the next stage of my life, during which I worked from a home office as one of the first fully independent microbiology consultants in the country.

With support from Sid Finegold, I became a coauthor of the sixth edition of *Bailey & Scott's Diagnostic Microbiology*. Writing that book became an avocation that continued through the ensuing twelve years. I discovered a joy of writing. I had the forum to present my ideas for diagnostic testing in a structured format—tempered of course by Sid's reasoned collaborative input. During the eighteen months of work required for each new edition, I never had to wonder what to do in the evenings or on the weekends.

From my early start in Wisconsin, I continued to present lectures and workshops. After one of my presentations, a local Los Angeles hospital pathologist called and asked if I would consult for his laboratory. I agreed and found consulting to be so much fun that I was embarrassed to accept payment. Through word of mouth, I soon found myself in such demand that I quit my Wadsworth VA job and made the plunge into becoming a self-employed microbiologist.

Because I had freedom to determine my own schedule, I was able to travel for relatively long periods. My international teaching activities blossomed. Dr. John Craig, Director of Pathology at one of my consulting hospitals, introduced me to Dr. Victor Lee, a founding father of Pathologists Overseas, which provides volunteer pathologists to tiny laboratories in exotic locations in the resource-poor world. Victor had just returned from a stint at Patan Hospital in Katmandu, Nepal, and he referred the staff there to me. I began a fax and e-mail correspondence with the missionary technologists running the lab there. When I was asked to speak at the Association of South East Asian Nations Society for Medical Laboratory Technology in Kuala Lumpur, Malaysia, the next year, two technologists from the Nepalese hospital attended the meeting. They sought me out and begged me to come to Patan to work with them in their laboratory.

As a volunteer at my own expense, I spent several weeks working on the bench with the local technologist and the missionary volunteer scientists at Patan Hospital. My eyes were opened to the challenges of doing quality microbiology with virtually no resources. Water wasn't clean enough to wash glass slides, the only paper was torn from catalog pages, and blood culture media was made in discarded whiskey bottles. Nevertheless, there was tremendous cooperation between the

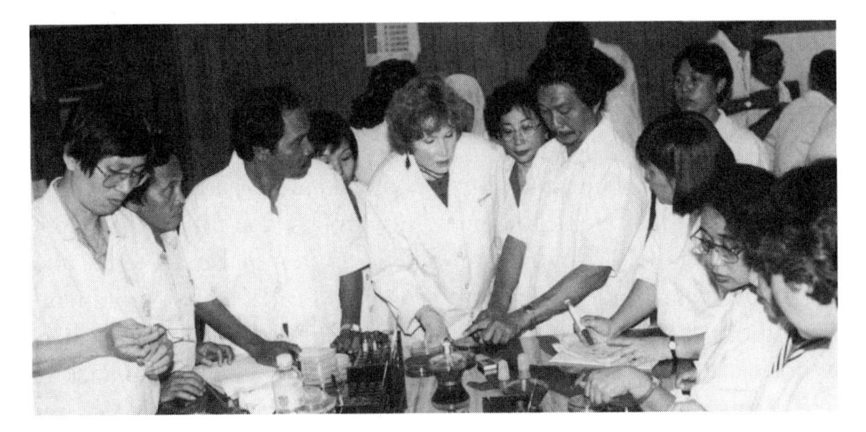

Teaching a class on basic bacteriology and detection of resistant organisms at my first WHO antimicrobial resistance monitoring course taught in Surabaya, Indonesia, in 1996.

laboratory and the physicians (also many volunteers along with the local doctors) and we made do!

After I returned, Tom O'Brien, codeveloper of WHONET software for susceptibility data analysis, invited me to become a trainer for a World Health Organization (WHO) course. And so my association with WHO's Antimicrobial Resistance Monitoring program began with a week-long workshop in Indonesia. The program has continued with other such workshops in remote parts of the world. My teaching experiences have been broadened and I am proud to say that I can bleed a sheep with the best of them now. I consider these visits to be among my most important contributions, as the microbiologists in these countries need so much help. I continue to work with WHO and other agencies to create training materials for resource-poor nations, many of whom are just beginning to develop their diagnostic microbiology capabilities.

My foreign travels, however, have been curtailed lately due to the demands of my latest position—Director of the Microbiology/Virology Laboratory at Stanford University Hospital. Recruited by one of my role models, Lucy Tompkins, I am privileged to continue my association with outstanding scientists, clinicians, students, and microbiologists. The demands of managing 50+ people and a $9 million budget, particularly in a time of tremendous fiscal crisis, have added to my career skills, which now include crisis management.

Speaking with companions during my travels and interacting with interested nonscientist friends and relatives have convinced me that few laypersons appreciate the value that microbiologists bring to everyone's lives through the practice of our science. The desire to spread the gospel of microbiology encouraged me to participate in high school career days, speak to community groups, and to join and work for the Association for Women in Science. What makes an active and successful woman microbiologist? Strong and early support from family, intense curiosity about the world around us, inspirational teachers at critical times, nurturing and sharing friends and colleagues, lack of fear of failure, and a hefty dose of both luck and perseverance.

I am extremely lucky, for I love microbiology—the detective work involved in identifying an unusual isolate or the satisfaction of helping a physician make a diagnosis or treat a patient. Every day brings new challenges, but it is still fun and rewarding at the same time. It is a gift to enjoy what you spend the most time doing, and to receive positive reinforcement and even compensation for doing it is even better. Just as I was helped at important stages of my development by others in the field, I try to help those around me to blossom and evolve into their potential best, in an effort to give back a portion of the gifts that I have received.

ELLEN JO BARON received her doctoral degree from the University of Wisconsin and served as a postdoctoral fellow at the University of California, Los Angeles, and the Wadsworth VA Medical Center. After serving as Chief of Clinical Microbiology at North Shore University Hospital in New York, she went on to become Director of the Clinical Anaerobic Bacteriology Research Laboratory at the Wadsworth VA Medical Center. She then became Consulting Director of Microbiology and Immunology at Endocrine Sciences in California, consultant to a number of hospitals in the Los Angeles area, and a member of the faculty at both the University of Southern California and the University of California, Los Angeles. She is currently Director of the Microbiology/Virology Laboratory at Stanford University Hospital and Associate Professor of Pathology, Stanford University Medical School.

The following books and papers are representative of her publications:

Baron, E. J., L. R. Peterson, and S. M. Finegold. 1994. *Bailey & Scott's Diagnostic Microbiology* (9th ed.). Mosby, St. Louis, Mo.

Murray, P. R., E. J. Baron, M. Pfaller, F. Tenover, and R. Yolken (ed.). 1999. *Manual of Clinical Microbiology* (7th ed.) ASM Press, Washington, D.C.

De la Maza, L. M., M. T. Pezzlo, and E. J. Baron. 1997. *Color Atlas of Diagnostic Microbiology.* Mosby, St. Louis, Mo.

Baron, E. J., G. Cassell, L. Duffy, D. Eschenbach, J. R. Greenwood, S. Harvey, and E. Peterson. 1993. Cumitech 17A, *Laboratory Diagnosis of Female Genital Tract Infections*, 2nd ed. In Cumitech Series, ASM Press, Washington, D.C.

Wenzel, R. P., D. R. Reagan, J. S. Bertino, E. J. Baron, and K. Arias. 1998. Methicillin-resistant *Staphylococcus aureus* outbreak: A consensus panel's definition and management guidelines. *Am. J. Infect. Control* **26:**102–110.

Baron, E. J., J. Downes, P. Summanen, M. Roberts, and S. M. Finegold. 1989. *Bilophila wadsworthia*, gen. nov., sp. nov., a unique gram-negative anaerobic rod recovered from appendicitis specimens and human faeces. *J. Gen. Microbiol.* **135:**3405–3411.

Baron, E. J., and D. M. Citron. 1997. Anaerobic identification flowchart using minimal laboratory resources. *Clin. Infect. Dis.* **25:**5143–5146.

Baron, E. J. 1995. Quality management and the clinical microbiology laboratory. *Diag. Microbiol. Infect. Dis.* **23:**23–34.

Tuner, K., E. J. Baron, P. Summanen, and S. M. Finegold. 1992. Cellular fatty acids in *Fusobacterium* species as a tool for identification. *J. Clin. Microbiol.* **30:**3225–3229.

Baron, E. J., R. Bennion, J. Thompson, C. Strong, P. Summanen, M. McTeague, and S. M. Finegold. 1992. A microbiological comparison between acute and complicated appendicitis. *Clin. Infect. Dis.* **14:**227–231.

D'Amato, R. F., H. D. Isenberg, G. A. McKinley, E. J. Baron, R. Tepper, and R. Shulman. 1988. Novel application of video image processing to biochemical and antimicrobial susceptibility testing. *J. Clin. Microbiol.* **26:**1492–1495.

Baron, E. J., and L. S. Young. 1986. Amikacin, ethambutol, and rifampin for treatment of disseminated *Mycobacterium avium-intracellulare* infections in patients with acquired immune deficiency syndrome. *Diag. Microbiol. Infect. Dis.* **5:**215–220.

COMBATING MICROBES

"Find the microbe—kill the microbe." With this outcry, Louis Pasteur—at least as portrayed by Paul Muni in his Academy Award–winning performance depicting the life of Pasteur—declared war on the microbes of death and disease. And so De Kruif says: "Do not think for a moment that Pasteur allowed his fame and name to be forgotten in the excitement kicked up by the sensational proofs of Koch that microbes murder men... Koch had just swept the German doctors off their feet by his fine discovery of the spores of anthrax." With four simple postulates, he had proved that microbes are our most deadly enemies. "And it was then that Pasteur... had the effrontery to dismiss the ten thousand years of experience of doctors in studying and fighting disease." The path was set for the conquest they would lead along with their legions—protégés they would individually train and the many followers who would march with them to combat those microbial menaces.

So it was that Ehrlich, Metchnikoff, Roux, and a horde of others joined the fray—the battle for the conquest of deadly diseases—for the survival of humankind. Paul Ehrlich was a physician turned microbiologist and biomedical researcher in search of cures—"a combative revolutionist... part of the revolt led by that chemist, Louis Pasteur, and the country doctor Robert Koch… declaring we must learn to shoot microbes with magic bullets." Elie Metchnikoff was a zealot, a mad Russian, a hysterical character from a Dostoevsky novel in search of immunity. Emil Behring was the "poetical pupil of Koch... bumping his nose in the darkness of his ignorance against facts the gods themselves could not have predicted." Emile Roux was "the financial helper of Pasteur… who dug ghoulishly into the spleens of dead children."

So it was that Ehrlich discovered—or rather stumbled onto—compound 606, an arsenical that could fight syphilis-causing bacteria. "It is evident from these experiments that, if a large enough dose is given, the spirochetes can be destroyed absolutely and immediately with a single injection!" So it was that Metchnikoff revealed the battle between phagocytes and marauding microbes. "'It is the phagocytes that eat up germs and so defend us,' roared Metchnikoff." So it was that Roux and Behring discovered the diphtheria antitoxin. "The serum of rats kills anthrax germs—it is the blood of animals not their phagocytes that makes them immune to microbes," yelled Emil Behring. And so it was, as De Kruif describes, that Louis Pasteur became the modern messiah, staging a miracle so grand that even his skeptics "bowed down before this man who could so perfectly protect living creatures from the deadly stings of sub-invisible invaders."

"Since the death of the child was almost certain," Pasteur wrote in 1885, "I decided in spite of my deep concern to try on Joseph Meister the method which had served me so well with dogs… I decided to give a total of 13 inoculations in ten days. Fewer inoculations would have been sufficient, but one will understand that I was extremely cautious in this first case. Joseph Meister escaped not only the rabies that he might have received from his bites, but also the rabies which I inoculated into him."

The mad dog no longer a threat, the pox no longer feared, the era of vaccines and antimicrobics at hand—so the roots of modern medicine arrived and with it the riches of longevity. Thus, when Fleming discovered penicillin and later Waksman streptomycin, and when Salk and Sabin took away the dread fear of polio, and when so many more wonder drugs were discovered in the last 50 years, it most certainly seemed that microbe hunters had won the war over those menacing invisible invaders of our bodies. It was the era of medical conquest. It was a time to celebrate, a time to declare victory over the microbes. Soon there would be no more infectious diseases—no pneumonia, no tuberculosis, none of a hundred other plagues that had afflicted humankind from the time of birth. Soon there would be no more pneumococcus, or tubercle bacteria, or any of the other bacteria, fungi, and viruses that Koch, Pasteur, and others of centuries gone by had sought to find and conquer. Soon there no longer would be a need for microbe hunters. A fairytale ending for our story.

But such was not to be the case. Alas, the microbe does seem to have the final word. New diseases like AIDS and Ebola emerged as plagues at the end of the twentieth century—as deadly as any ever recorded in the history of humankind. Old nemeses like tuberculosis reemerged, and, worse yet, bacteria evolved resistance to penicillin and almost all of the other "magic bullets." We need a new generation of microbe hunters—those brave enough to seek cures for a myriad of infectious diseases. There are new plagues to conquer, new magic bullets to find, and a new millennium for seeking cures to human disease. Stand up, ye new microbe hunters. Stand up and fight. Find the microbes—and kill them dead.

Immune Legions Battling Infection

Humankind has suffered many afflictions—some from the plagues of invisible microbes, some from natural disasters, and some from man's inhumanity to man. I am a first-generation American. My parents were immigrants from Eastern Europe who left Russia to escape pogroms, especially against poor villagers in the "pale" of Russia, which consisted of sections of what is now Poland and the Ukraine. They were among the hordes that escaped from Eastern Europe in the 1920s after World War I. It was my fortune that they made it to the United States before the outbreak of World War II. Their bold move prevented their extermination, which was the fate of most Jews who remained in Russia and Eastern Europe during the Holocaust. We survived.

As immigrants, my parents did not know English and lived in a poor area in North Philadelphia. I was the second of three children and the only one to live to adulthood. I did not learn English until I went to grammar school. Like most of my other young relatives and friends in the neighborhood, my first language was Yiddish. My memories of going to a public school in that poor North Philadelphia neighborhood are still very vivid. Like many immigrant children, doing well in school was a high priority for me. I went to Central High School, which was and is the premier public high school in Philadelphia. It was founded in 1836 as only the second public high school in America and chartered in Pennsylvania to grant bachelor of science degrees. Until World War II, graduates from Central were accepted directly into medical or other professional schools without going to an undergraduate college.

I graduated from Central and went on to Temple University. It was considered the "poor man's college" in Philadelphia and had extremely low tuition—less than just the fees at the University of Pennsylvania. I majored in biology, and the last two years I took microbiology courses. My microbiology professor, Dr. James Harrison, fostered in me a curiosity about how humans could live surrounded by a sea of pathogenic microorganisms. Thus, I began my lifelong interest in how the immune system developed, how it provided resistance against microbes, how autoimmune diseases develop, and how diseases such as cancer occur when the immune system fails.

First in high school, then at Temple, my major interest was journalism. I was editor-in-chief of the high school newspaper, which consistently won national prizes, including the Columbia University Journalism prize for secondary schools. At Temple I became involved in the *Temple News*, which was a daily paper, and became editor-in-chief. The faculty of the journalism department was astonished that I was a biology major. The science faculty, including Dr. Harrison, thought that editing the paper was a waste of my time—time that should have been devoted to my study of science. My view then, as it is now, is that writing and science are two sides of the same coin. Communicating is an essential part of being a scientist.

When I finished my undergraduate degree at Temple, I began working in the research laboratories of Zvi Harris at the Children's Hospital of Philadelphia at the University of Pennsylvania and was enrolled for a master's degree in the biology program at Temple. Harris, in the Pediatrics Department at the University of Pennsylvania School of Medicine, permitted me to use my research experiments in his laboratory for my master's thesis at Temple. Harris was a pioneer in the study of the immune response to bacteria. He was the first immunologist to propose that lymphocytes, which were considered then to be an end-stage cell without any known function, were the source of antibodies. In doing so, Harris had broken ranks with George Ehrlich, a renowned professor at the University of Pennsylvania who supported the concept that only plasma cells produced antibodies.

When I joined Harris's lab as a "wet behind the ears" newly minted college graduate of 21 years of age, I very excitedly told many of my friends that Harris was doing miraculous experiments—lethally irradiated rabbits could be made to respond to *Shigella* vaccine after restoring their immune response system by merely transferring into the animals popliteal lymph node cells from normal donor rabbits. As a technician, I did experiments that also showed that lymphoid cells from a rabbit could produce antibody in vitro when stimulated with *Shigella* vaccine in a test tube. This was unheard of at that time. The Harris lab was one of the first to show that lymphocytes were involved in the immune response to any antigen, including bacteria that were known to cause disease.

Dr. Harris' laboratory at the University of Pennsylvania was called the Rheumatic Fever Research Laboratory. His laboratory was mainly responsible for studying the role of streptococci in rheumatic fever. For that work, one of my duties was to purify streptococcal enzymes thought to be involved in rheumatic fever pathogenesis. I used those results for my thesis at Temple University and completed the master's degree in two years. I then began a Ph.D. program at Hahnemann Medical College. I was only the second student in the microbiology department to enter the program. At that time very few students even thought of leaving their home city for college or graduate work. In Philadelphia there were only four Ph.D. programs in microbiology that interested me, and I was accepted at all. When the microbiology department at Hahnemann Medical College offered free tuition and a generous stipend of $2,000 per year, I quickly accepted.

I began my graduate studies at Hahnemann studying mainly biochemical aspects of purine metabolism in yeast, but I also continued some of my own experiments on the immune responses to *Shigella* in rabbits. I received my Ph.D. after two years of study there and applied for my first National Institutes of Health (NIH) grant two years later at age 27 with one of the faculty members at Hahnemann as coinvestigator. The grant was rejected, which was quite rare in the mid-1950s, because even though it had only $100 million to support extramural programs, NIH funded about 90% of all grant applications; the real purpose of study sections was to recommend which 10% of the grants should not be supported. The program director at NIH referred me to one of the reviewers on the study section, who told me that simply changing one thing throughout the grant would guarantee funding. The change was to replace "*Shigella paradysenteriae*" with "bovine serum albumin," since "all

Faculty of the Department of Medical Microbiology and Immunology, University of South Florida College of Medicine. (L to R): Steven Specter (Professor), Susan Pross (Assistant Professor), Burt Anderson (Associate Professor), Herman Friedman (Professor and Chair), Allan Honeyman (Assistant Professor), Peter Medveczky (Professor), Thomas Klein (Professor), Kenneth Ugen (Assistant Professor).

immunologists knew that only the study of pure antigens was of value—no one could really study immune mechanisms using bacteria or microorganisms." I refused, and it took me a decade to finally receive funding from NIH to study the immune response to microbial antigens. Fortunately, my funding from NIH has continued since then for the past three decades. Many immunologists have followed my lead and now study immune responses to bacteria, viruses, and protozoa, rather than pure antigens—a focus seen in today's research on preventing and curing AIDS and a variety of other infectious diseases.

This brings me to my laboratory career. My first full-time professional position was as the director of an allergy research laboratory at the Veterans Administration Hospital in Pittsburgh. I soon recognized that research work in a government laboratory did not interest me. So I returned to Philadelphia and became Director of the Clinical Microbiology and Immunology Laboratory at Albert Einstein Medical Center and a faculty member at Temple University School of Medicine. I was fortunate in having a relatively large number of graduate students and many postdoctoral research associates. I continued with my studies on basic mechanisms of immune resistance and tolerance to bacteria, including *Shigella*, *Vibrio cholerae*, and *Salmonella*.

In the mid-1960s, I began studying the effect of Friend leukemia virus, now known to be a retrovirus, on the immune response by investigating the role of subcellular RNA factors in transferring immunity to microorganisms. Animals infected with this leukemia virus became susceptible to a variety of infections by opportunistic bacteria, such as *Escherichia coli* and pseudomonads. We found that mice infected with the virus became immunosuppressed and showed a defect in their antibody response to bacteria and sheep erythrocytes as well as a deficiency of what we now know is a T cell–mediated mitogenic response. Most of my colleagues downplayed the importance of this finding, but fortunately we persisted, because it turned out that the way this retrovirus induced immunosuppression was analogous to the way HIV causes AIDS. Thus, years before the AIDS epidemic, we were studying how a virus could induce immunosuppression and how it could increase susceptibility to infections by opportunistic pathogens.

In 1976, while I still was Director of the Diagnostic Microbiology Laboratory at Einstein Medical Center, there was a major outbreak of severe pneumonia. The disease was called *Legionnaires' disease* because the outbreak occurred mainly among those who attended a Legionnaires' convention at the Bellevue Stratford Hotel. My laboratory at Einstein investigated the role of the immune response in host susceptibility and resistance to this organism, both at the serum level and at the level of their blood lymphoid cells. After I relocated to Florida, we continued studies on Legionnaires' disease that we had begun in Philadelphia. We used *L. pneumophila* as a model system and studied immune responses of mice to this infectious organism. We developed a susceptible versus resistant mouse model to study *Legionella* and investigated the role of B cells, T cells, and macrophages, as well as cytokine responses to infection by this bacteria. Thus, this investigation started initially in a clinical laboratory but continued in a research laboratory.

In addition to studying the immune response to *Legionella*, we soon began to examine the effect of marijuana on the immune response system. These studies originated with the recognition that drug abusers were especially prone to HIV infection and the development of AIDS. Our studies showed that animals treated with tetrahydrocannabinol, the active substance in marijuana, became markedly immunosuppressed and much more susceptible to infection with opportunistic pathogens, such as *E. coli*, *Legionella*, and *Candida albicans*. Our studies on immunosuppression and illicit drug use, which were initiated in the clinical laboratory for practical needs, soon developed into basic research investigations.

As told by Dr. Harrison at Temple University over 40 years ago, if one performs good science in an important area of study, the world will eventually come around; we should continue with our own interests and not jump into areas of study simply because they are in vogue. Most biomedical scientists accept the fact that we have a long way to go before we understand the nature and mechanism of host resistance to infectious diseases. Nevertheless, it is important to realize that over the last several decades, many advances have been made in molecular biology, physics, and chemistry and that the knowledge so gained can be used to now study in more detail and with more useful methods host resistance mechanisms to microorganisms. We are in the midst of the most exciting era to study immune mechanisms against microbes. I believe that the next few decades will result in a better understanding of the host-parasite relationship and also continue major advances that translate into treatment for and cure of many infectious diseases that have been the scourge of humans for eons.

One of the stories I was told many times as a child was that my father's father died in Russia about 1905 because of an infection that was lethal at that time for anyone whether he or she was the tsar or president or the richest person in the world—peritonitis caused by a ruptured appendix. It was only after the antibiotic era became a reality in the 1940s and 1950s that we were no longer threatened with certain death by such infections. Indeed, many infectious diseases have been controlled by the judicious use of antibiotics. However, we now know that overuse and misuse of antibiotics have resulted in emergence of many antibiotic-resistant organisms that we thought were easily controllable. Other serious infections by emerging pathogens—be they viruses, parasites, or fungi—occur because of the increase in the number of individuals who have either innate or acquired immunodeficiency. Only by achieving a greater understanding of the nature and mechanism of immune responses to microbes will further control of infectious diseases by microorganisms be possible. The last 50 years have certainly shown a remarkable increase in our understanding of host immunity to microbes. It is obvious that further investigation of the host-parasite relationship as it relates to immune responses against microorganisms will result in what was thought to be an unattainable understanding of how the human species can survive in the "sea of microbes" that can use humans as their hosts.

HERMAN FRIEDMAN received his doctoral degree from MCP Hahnemann University School of Medicine, where he also completed his postdoctoral training. For nineteen years he was head of the Department of Microbiology and Immunology at Albert Einstein Medical Center in Philadelphia. During that time he began his research on immunity and microbial infection. In 1978, he moved to the University of South Florida, where he is distinguished professor and chair of the Department of Medical Microbiology and Immunology. His studies of the effects of viruses, bacteria, and their products on immune system components have made major contributions in many areas of immunology. He has made significant contributions to the field of cellular immunology during the past forty years—in particular, by revealing mechanisms of antibody formation, immunologic tolerance, and immunosuppression. His accomplishments have been recognized with the following awards: Distinguished Professor, University of South Florida, 1994; Becton Dickinson Clinical Microbiology Award, 1987; American Society for Microbiology Distinguished Service Award, 1996; and the Abbott Laboratory Award in Clinical and Diagnostic Immunology, 1999.

The following papers are representative of his publications:

Friedman, H. 1996. Drugs of abuse as co-factors in AIDS progression. *Exper. Med. Biol.* **405:**225–229.

Yamamoto, Y., T. W. Klein, C. Newton, and H. Friedman. 1992. Differing macrophage and lymphocyte role in resistance to *L. pneumophila*. *J. Immunol.* **148:**584–589.

Friedman, H., and A. Szentivanyi. 1985. Antibacterial immunity, vaccines, and allergy. *Allergology* **8:**357–369.

Friedman, H. 1973. Cellular and molecular aspects of immune responses to bacterial somatic antigens. *J. Infect. Dis.* **128:**561–569.

Ceglowski, W., and H. Friedman. 1969. Murine virus leukemogenesis: relationship between susceptibility and immunosuppression. *Nature* **224:**1318–1321.

Friedman, H. 1968. Effect of serum from tolerant mice on immunity and tolerance to a bacterial antigen. *Nature* **218:**1261–1268.

Friedman, H. 1965. Failure of spleen cells from immunologically tolerant mice to form antibody plaques to sheep erythrocyte in agar gel. *Nature* **205:**508–510.

Friedman, H. 1964. Antibody plaque formation by normal spleen cell cultures exposed *in vitro* to RNA from immune mice. *Science* **146:**934–938.

Friedman, H. 1964. Inhibition of antibody plaque formation by sensitized lymphoid cells: rapid indicator of transplantation immunity. *Science* **145:**607–609.

Friedman, H. 1962. Transfer of antibody formation by spleen cells from immunologically unresponsive mice. *J. Immunol.* **89:**257–269.

Pursuing Drug-Resistant Bacteria

My father was a practicing physician in Wilmington, Delaware. In those days physicians still made house calls, including on Sundays. Some of my fondest and earliest childhood images are of those Sunday mornings—riding in the car with my identical twin brother, Jay—seeing my father caring for the ill. On those outings, we witnessed the respect my father received from his patients; some were only able to pay with food or services in return. It was wonderful to see the care my father gave them. My brother and I would both become physicians and researchers—albeit on different microbes and on opposite sides of the country.

But although my interest in medicine was engendered on those early Sunday morning drives, I had an even greater fascination with nature—catching animals, collecting insects, and growing trees and plants on the land next to our house. My early interests in science and medicine became a long-lasting part of my life. As a student at Williams College, I decided not to major in science. Instead I chose English. I loved literature, and I wanted to give that interest some rein before I made any decision about my future career. I toyed between careers in medicine and law. Along with courses in art, music, and other liberal arts, I took the science courses required for entry into medical school. In many ways, this strategy embodied how I led and lead my life: keep your options open. Only make trade-off decisions when necessary.

By the end of my third year at college, I had decided to go to medical school. I chose the University of Pennsylvania, whose excellent reputation was coupled with attention to the needs and desires of each student. I was not wrong in that decision; I was given considerable breadth in my medical education. Penn provided first-year students with the opportunity to perform bench research with a professor. I was accepted by Joseph Gots, Professor of Microbiology, well known in the field of bacterial genetics. My experience in his laboratory with his postdoctoral fellows and students stimulated and solidified my burgeoning interest in research. At the end of my first year, I took advantage of a university summer fellowship to do research on bacteriophages, under the tutelage of Giulio Maccacaro at the Istituto di Microbiologia in Milan. There, coincident with research in microbiology, I also learned Italian and developed a love for Italy—rewards from the experience that remain with me today.

The excitement of being on my own in research in Italy groomed me for other research experiences. I spent the second year of medical school figuring out how I could take a year off for additional research before beginning the clinical rotations. I sought a foreign country so I could combine science with culture. I think I was reacting in part to the rather insular and provincial environments of Wilmington and Williams. I chose Paris as my city, and after persistent correspondence to known leaders in the bacterial genetics field, I was delighted to receive a two-line acceptance from Raymond Latarjet at the Pasteur Institute in Paris. With Raymond, who became both mentor and friend, I worked independently in the area of radiation genetics. Bacteria contacted with x-ray–inactivated bacteriophages (well known to me after my summer in Milan) were resistant to live phage infection. These findings in bacteria served as a model for stimulating resistance to mammalian cell tumor viruses being performed in Latarjet's group. That year's experience changed my life enormously.

I learned French and French culture; this certainly helped in my eventual marriage to a Parisian physician. It led to my first publication in the *Journal of Bacteriology*, the professional journal of the American Society for Microbiology, an organization of which I would later be president. Moreover, at the time, I met a visiting Japanese scientist who introduced me, at journal club, to the fascinating transferable antibiotic resistance factors described in a publication by Tsutomu Watanabe in Japan.

On returning to Penn, in between my required, but not always full, attention to clinical rotations, I corresponded with Watanabe, hoping to be accepted for a few months in his laboratory the following fall. Having been back from Europe less than a year, I was already planning another break from medical school—testimony to Penn's flexibility. I was fortunate to be accepted and to receive fellowship funds. In fact, the summer and fall after my third year involved two back-to-back experiences in areas that eventually came together in my current career.

In the summer, on a Publiker Nutrition fellowship, I went to Kenya and did the first analysis of hemoglobin levels among eight African tribes. While I was there, I met some wonderfully supportive hematologists, Henry Foy and Athena Kondi, who became mentors and friends and introduced me to Dr. Munoz of UNESCO, from whom I received my own Land Rover for the study—quite an unusual experience for a third-year medical student. I came with the necessary reagents and advice from David Drabkin, my biochemistry professor at Penn, who provided me with his hemoglobin-determining solution. How lucky can you get! I visited a different tribe each week for eight weeks, living with them and having a translator to help obtain blood and stool samples, which I analyzed back in Nairobi. The studies were the first to demonstrate differences in hemoglobin values among different tribes of Kenya, reflecting altitude and the prevalence of hookworm and malaria.

From Kenya, I went to Watanabe's laboratory in Tokyo, where in the heat of September, we worked beside each other (wearing undershirts) in the relatively new area of transferable drug resistance. With Watanabe, I published two papers—one in Japanese, one in English—on methods of curing R plasmids with acridine dyes. I continued to work on transferable R factors in Joe Gots' laboratory on my return to the University of Pennsylvania for my final year of medical school.

These experiences in medical school helped me decide on an academic career that would combine the practice of medicine—which I did not want to leave—with basic research. I chose my medical residency—internship and first year—at an institution with broad clinical and research expertise: Mt. Sinai Hospital in New York, under Alexander Gutman. There, through Latarjet, I met Charlotte Friend, noted virologist and discoverer of the Friend erythroleukemia virus, in whose laboratory I spent time as a resident. Charlotte called me her "first graduate student"; I was charmed, as were many, by her warmth and attention and her love of research. We became best friends, seeing each other often until her untimely death. This experience introduced me to new techniques, namely tissue culture and tumor viruses, and spawned my interests in cell biology, hematology, and malignancy, which I carry today.

From Mt. Sinai, I became a Staff Associate in the laboratory of Loretta Leive at the National Institutes of Health (NIH) with the Public Health Service. I returned to microbiology, examining bacterial outer membrane lipopolysaccharide synthesis. Through Loretta, I experienced true biochemistry. Although I enjoyed the work, I was eager to return to antibiotic resistance.

While at the NIH, I learned about the fascinating *E. coli* mutant derived by Howard Adler at Oak Ridge, Tennessee, which spewed off tiny chromosomeless cells called *minicells*. They appeared to be the perfect means for identifying products specified by R plasmids. In minicells, the only source of DNA would be the plasmid; the chromosomal DNA dominated whole cells. I extended my NIH stay for a third year, and I put my idea to the test. To my delight, not only did the plasmid I chose, R222 (which I had brought back with me from Watanabe's laboratory), efficiently segregate into the minicell, but it brought with it the appropriate enzymes by which to make products that we could identify. In fact, it was the R222 plasmid-in-minicell system on which I continued working when I went for my fellowship training to Tufts-New England Medical Center in Boston the following year.

Although accepted in a clinical fellowship in hematology (under William Crosby), I wanted to pursue my interest in R factors. Through Joe Gots, I met Moselio Schaechter, Chair of the Department of Molecular Biology and Microbiology at Tufts, who provided space in his laboratory to continue my R factor studies. I also arranged time to make clinical rounds with Louis Weinstein in order to maintain contact with infectious diseases. My two interests in microbiology and hematology were united.

During my hematology fellowship, I followed up some research that I had begun with Charlotte Friend, placing leukemic spleens subcutaneously back into leukemic mice and noting prolongation of their lives. In retrospect, the concept of using biological material to reverse a disease process echoed my phage inhibition studies with Latarjet in Paris. It was also during this fellowship that I joined my brother under the auspices of Fred Prince of the New York Blood Center to study hepatitis B virus among wild baboons in the brush of the Ivory Coast. We had piloted this study the summer before in Uganda, where we found evidence for a primate Epstein-Barr virus in wild chimpanzees. The experiences were unique, combining ecology, nature, and microbiology.

The next year, through NIH support, I continued my studies of R plasmids at Tufts University School of Medicine, while maintaining my clinical responsibilities with the hematology division at the hospital. Together with my long-term colleague, Laura McMurry, we used minicells to discover the inner membrane Tet protein associated with tetracycline resistance. Later, we demonstrated that it was involved in active efflux of the antibiotic. In the late 1970s, this was the first demonstration of active efflux as a mechanism of drug resistance. This fact seems curious today when, almost weekly, reports of efflux systems for all kinds of antimicrobials and other substrates appear in the literature. Some time after our report, Vickers Burdett described a second mechanism for resistance to tetracycline—that is, protection of the ribosome. In fact, efflux and ribosome protection traits carried on acquired genes form the basis for all the major tetracycline resistances found in bacteria.

My laboratory then showed that the tetracycline resistance determinants were heterogeneous, distinguishable by DNA-DNA hybridization and later by sequencing. We established a nomenclature for these determinants based initially on the letters of the alphabet and more recently by Arabic numbers. We continued our in-depth studies of tetracycline resistance with the purification of the Tet protein and early studies of its molecular characteristics, both as purified protein in detergent and in crystal-growing studies with the expertise of Mila Aldema-Ramos in my laboratory and those of my collaborators.

Stemming from my experience with Watanabe, I also investigated the ecology of drug resistance and resistance plasmids. Setting up a small chicken farm outside of Boston, my research team was the first to demonstrate prospectively the transfer of antibiotic resistance from animals to humans. We also documented the selection by low-dose tetracyclines in animal feed of high-level tetracycline resistance in bacteria associated with chickens and their caretakers. We continued microbial ecology studies under the expert direction of Bonnie Marshall in my laboratory, showing the spread of bacteria among animals, flies, and people on a farm. Today, the widespread use of antibiotics in animals has become controversial as the explosive rise in antibiotic resistance poses a serious threat to medical uses of antibiotics to treat human diseases.

In a study in Amboselli Park, Kenya, performed by Rosalind Rolland, a Tufts veterinary student, with Bonnie Marshall and Glenn Hausfater (from Cornell), we showed how antibiotic-resistant intestinal bacteria from people were picked up by baboons rummaging through tourist refuse pits. These kinds of studies continue to be a source of exciting and interesting findings that all add to a greater understanding of the broad ecology of antibiotic-resistant bacteria and how antibiotic use affects not only pathogenic organisms but also the larger reservoir of commensal bacteria.

In attempts to look for a chromosomal mutant to tetracycline, I stumbled on an unusual locus in *Escherichia coli*, called *mar*, which specified *m*ultiple *a*ntibiotic *r*esistance. Consisting of four proteins and two operons, it controls how the cell senses the environment, causing a change in expression of more than sixty different genes residing elsewhere on the chromosome. With the *mar* locus activated by contact with multiple structurally unrelated chemicals, including aspirin, or by mutation, the cells pump out different kinds of antibiotics, detoxify toxic substances, and provide resistance to organic solvents, disinfectants, and surface antibacterials. These studies, like those previously described, have led to an appreciation of the intrinsic means by which bacteria respond to and can survive diverse unrelated environmental hazards.

Discovery of the *mar* locus led to our recent demonstration that antibacterials contained in household products are subject to a *mar*-mediated efflux system. We identified the cellular enoyl reductase as the target for the "biocide" triclosan—this enzyme is also a site of action for isoniazid in *Mycobacterium tuberculosis*. These findings have heightened the concern that overuse of antibacterial-containing household items will foster propagation of drug-resistant bacteria and impact the antibiotic resistance problem.

To bring the tetracycline resistance findings forth to benefit public health, I began to think of ways to curtail or circumvent tetracycline resistance. I decided to design decoy molecules based on tetracycline that block the resistance mechanisms and restore antibacterial activity to the standard tetracyclines. Mark Nelson joined me as the chemist, and together we developed a number of novel tetracyclines that were successful blockers of resistance, not only of efflux but also of ribosome protection.

To further turn the tetracycline and *mar* findings in my laboratory into available products for treatment of infectious diseases, I formed a company, Paratek Pharmaceuticals, Inc., with Wally Gilbert. One of the company's projects involves the renaissance of tetracyclines through the discovery and synthesis of derivatives active against resistant bacteria. Another focuses on inactivating the *mar* system as a means of enfeebling infecting organisms. While I remain on the faculty at Tufts University, I continue to help direct the science at Paratek.

Concerned about the continued misuse of antibiotics leading to the resistance problem, I organized a meeting in Santo Domingo, Dominican Republic, in 1981 with Roy Clowes and my sister, Ellen Koenig, who is a professor of microbiology and medicine there. Experts in infectious diseases, microbiology, and plasmid biology attended the sessions. We prepared an Antibiotic Misuse Statement. The statement was widely discussed and presented at press conferences simultaneously in four different countries and later translated into many different languages.

The message was, and still is, that there is a global public health problem stemming from the way we use antibiotics. Clearly antibiotics have revolutionized our ability to curb death and disease from infectious microorganisms. However, the seemingly endless miracles attributed to these drugs have led to their misuse and overuse. Antibiotic resistance emerges, limiting the effectiveness of these valuable therapeutic agents.

Bacteria have responded to the widespread applications of antibiotics by finding ways to become resistant, insensitive to the killing effects of these powerful drugs. The overuse of antibacterial agents kills off susceptible bacteria, enabling competitor flora in the environment to proliferate and cause infection, as well as causing some bacteria to actually develop mutations conferring resistance. Thus, antibiotics sow the seeds of their own potential downfall by selecting for rare strains of bacteria that have the ability to resist their activity. Many of these resistance traits can be transferred or spread from resistant bacteria to other bacteria, even of different types.

The overwhelmingly positive response to this message was pressed forward by our founding of the Alliance for the Prudent Use of Antibiotics (APUA). APUA seeks to involve the international community in the antibiotic resistance problem—something that had not been done before. Our mission is to raise awareness of the antibiotic resistance problem and promote proper use of antibiotics worldwide.

Having lectured extensively to physicians and health providers, I decided to take the message to the consumers, whose demanding and stockpiling of these drugs were, to me, prominent contributors to the problem. I wrote a book called *The Antibiotic Paradox: How Miracle Drugs Are Destroying the Miracle*. In it, I discuss the consequences of antibiotic misuse, highlighting the many resistance problems worldwide at that time. Today, the situation continues to be of great concern. Some isolates of at least five different organisms are nontreatable, and we have witnessed the emergence of vancomycin-resistant *Staphylococcus aureus*. Most recently, community-acquired methicillin-resistant *S. aureus* (MRSA) strains have appeared, resulting in the deaths of four children in the United States. The problem is complicated because no structurally new antibiotics are scheduled for release in the near future.

It is somewhat curious how fate has played a role in where I am today, along with my aggressive pursuit of what I found interesting. Many individuals steered me onto the path of bacteria, while others showed me the attraction of hematology and cell biology. Ironically, my twin brother, Jay, who also chose academic medicine, focused on infectious diseases and cancer, devoting his attention largely to viruses. Although remaining close, we settled on different coasts. Today I see hospitalized patients as part of the Hematology/Oncology Division at the New England Medical Center, I direct a research center at the Tufts University School of Medicine, serve as Chief Scientific Officer of Paratek Pharmaceuticals, and continue as President of APUA. These positions and activities satisfy my quest and my need to see a broader role for this research in public health. Together they incorporate the many different aspects and roots of my professional career.

In sum, the leading influence on my life has been interesting science, coupled with my refusal to be pegged into a particular niche. I always saw the link between bacterial and mammalian cells, from the early days with Latarjet and Friend, to the important influence of Watanabe. These two scientific areas now have come together in my Center for Adaptation Genetics and Drug Resistance at Tufts, where the parallels are emphasized and where the training of fellows includes both prokaryotic and eukaryotic systems. Looking back on where I am today—a physician-scientist and microbiologist involved in the academic, clinical, public health, and business worlds—it is personally intriguing and enlightening to piece together those people and events which influenced my decisions that led me here.

I have become a voice for the proper use of antibiotics. I carry the message around the globe that we must make peace with the bacteria. We entered their world, not vice versa. We have already misused antibiotics and, more recently, antibacterial-laced household products under the false impression that we can or should somehow rid the world of bacteria—a feat we cannot achieve, and if we did, that would lead to our own demise.

STUART B. LEVY is Professor of Molecular Biology and Microbiology and of Medicine and the Director of the Center for Adaptation Genetics and Drug Resistance at Tufts University School of Medicine. He is also Staff Physician at the New England Medical Center. Born in Wilmington, Delaware, he is a magna cum laude, Phi Beta Kappa graduate of Williams College. Dr. Levy received his medical degree from the University of Pennsylvania. He did his residency at Mt. Sinai Hospital in New York and postdoctoral research at the National Institutes of Health. He held research fellowships in Paris, Milan, and Tokyo. He serves as President of the Alliance for the Prudent Use of Antibiotics. He is a Fellow of the American College of Physicians, the Infectious Disease Society of America, and the American Academy of Microbiology. He was awarded the 1995 Hoechst-Roussel Award for esteemed research in antimicrobial chemotherapy by the American Society for Microbiology (ASM). In 1998, he was awarded an honorary degree in biology from Wesleyan University in Middletown, Connecticut. He was the 1998–1999 President of ASM. In 1992, he established and became director of the Center for Adaptation Genetics and Drug Resistance at Tufts University School of Medicine, where he conducts research in the areas of resistance to antibiotics and to anticancer drugs. He has been featured and quoted for his work on antibiotic use and resistance in major national and international newspapers and magazines, and he has appeared on National Public Radio and all major American television network news shows, as well as on Canadian national television and Japanese public television. He was featured in the March 28, 1994, *Newsweek* magazine cover story on antibiotic resistance and in the lead article on antibiotic resistance in *Discover* magazine in November 1998. In his work, he has had the opportunity to mentor dozens of aspiring young scientists, with both M.D.s and Ph.D.s. Many of these individuals have gone on to establish themselves in outstanding research careers.

The following books and papers are representative of his publications:

Alekshun, M. N., and S. B. Levy. 1999. Characterization of MarR superrepressor mutants. *J. Bacteriol.* **181:**3303–3306.

Levy, S. B., L. M. McMurry, T. M. Barbosa, V. Burdett, P. Courvalin, W. Hillen, M. C. Roberts, J. I. Rood, and D. E. Taylor. 1999. Nomenclature for new tetracycline resistance determinants. *Antimicrob. Agents Chemother.* **43:**1523–1524.

Nelson, M. L., and S. B. Levy. 1999. Reversal of tetracycline resistance mediated by different bacterial tetracycline resistance determinants by an inhibitor of the Tet(B) antiport protein. *Antimicrob. Agents Chemother.* **43:**1719–1724.

Levy, S. B. 1998. Multidrug resistance: a sign of the times. *N. Engl. J. Med.* **338:**1376–1378.

McMurry, L. M., M. Oethinger, and S. B. Levy. 1998. Triclosan targets lipid synthesis. *Nature* **394:**531–532.

Levy, S. B. 1998. The challenge of antibiotic resistance. *Sci. Am.* **278:**46–53.

Moken, M. C., L. M. McMurry, and S. B. Levy. 1997. Selection of multiple antibiotic resistant (Mar) mutants of *Escherichia coli* by using the disinfectant pine oil: roles of the *mar* and *acrAB* loci. *Antimicrob. Agents Chemother.* **41:**2770–2772.

Alekshun, M., and S. B. Levy. 1997. Regulation of chromosomally mediated multiple antibiotic resistance: the *mar* regulon. *Antimicrob. Agents Chemother.* **41:**2067–2075.

Draper, M. P., R. L. Martell, and S. B. Levy. 1997. Active efflux of the free acid form of the fluorescent dye 2',7'-bis (2-carboxyethyl)-5(6)-carboxyfluorescein in multidrug-resistance-protein overexpressing murine and human leukemia cells. *Eur. J. Biochem.* **243:**219–224.

Modrak, D. E., M. P. Draper, and S. B. Levy. 1997. Emergence of different mechanisms of resistance in the evolution of multidrug resistance in murine erythroleukemia cell lines. *Biochem. Pharmacol.* **54:**1297–1306.

Aldema, M. L., L. M. McMurry, A. R. Walmsley, and S. B. Levy. 1996. Purification of the Tn*10*-specified tetracycline efflux antiporter TetA in a native state as a polyhistidine fusion protein. *Molec. Microb.* **19:**187–195.

Levy, S. B. 1995. Antimicrobial resistance: a global perspective. *Adv. Exper. Med. Biol.* **390:**1–13.

Nelson, M. L., B. H. Park, and S. B. Levy. 1994. Molecular requirements for the inhibition of the tetracycline antiport protein and the effect of potent inhibitors on growth of tetracycline resistant bacteria. *J. Med. Chem.* **37:**1355–1361.

Slapak, C. A., P. M. Fracasso, R. L. Martell, D. L. Toppmeyer, J.-M. Lecerf, and S. B. Levy. 1994. Overexpression of the multidrug resistance associated protein (MRP) gene in vincristine but not doxorubicin-selected multidrug resistant murine erythroleukemia cells. *Cancer Res.* **54:**5607–5613.

Levy, S. B. 1993. Confronting multidrug resistance. A role for each of us. *JAMA* **269:**1840–1842.

Cohen, S. P., H. Hachler, and S. B. Levy. 1993. Genetic and functional analysis of the multiple antibiotic resistance (*mar*) locus in *Escherichia coli*. *J. Bacteriol.* **175:**1484–1492.

Cohen, S. P., W. Yan, and S. B. Levy. 1993. A multidrug resistance regulatory chromosomal locus is widespread among enteric bacteria. *J. Infect. Dis.* **168:**484–488.

Levy, S. B. 1992. *The Antibiotic Paradox. How Miracle Drugs Are Destroying the Miracle*. Plenum, New York, N.Y.

Levy, S. B. 1992. Active efflux mechanisms for antimicrobial resistance. *Antimicrob. Agents Chemother.* **36:**695–703.

Levy, S. B. 1990. Starting life resistance-free. *N. Engl. J. Med.* **323:**335–337.

Rubin, R. A., S. B. Levy, R. L. Heinrikson, and F. J. Kezdy. 1990. Gene duplication in the evolution of the two complementing domains of gram-negative tetracycline efflux proteins. *Gene* **87:**7–13.

Marshall, B. M., D. Petrowski, and S. B. Levy. 1990. Inter- and intraspecies spread of *E. coli* in a farm environment in the absence of antibiotic usage. *Proc. Natl. Acad. Sci USA* **87:**6609–6613.

Cohen, S. P., D. C. Hooper, J. S. Wolfson, K. S. Souza, L. M. McMurry, and S. B. Levy. 1988. Endogenous active efflux of norfloxacin in susceptible *Escherichia coli*. *Antimicrob. Agents Chemother.* **32:**1187–1191.

Levy, S. B., B. Marshall, S. Schluederberg, D. Rowse, and J. Davis. 1988. High frequency of antimicrobial resistance in human fecal flora. *Antimicrob. Agents Chemother.* **32:**1801–1806.

Levy, S. B., and R. P. Novick. 1986. *Antibiotic Resistance Genes: Ecology, Transfer, and Expression*. Cold Spring Harbor Laboratory, Cold Spring Harbor, NY.

Rolland, R., G. Hausfater, B. Marshall, and S. B. Levy. 1985. Increased frequency of antibiotic resistant gut flora in wild baboons in contact with human refuse. *Appl. Environ. Microbiol.* **49:**791–794.

George, A. M., and S. B. Levy. 1983. A gene in the major co-transductional gap of the *Escherichia coli* K-12 linkage map required for the expression of chromosomal resistance to tetracycline, chloramphenicol and other antibiotics. *J. Bacteriol.* **155:**541–548.

Mendez, B., C. Tachibana, and S. B. Levy. 1980. Heterogeneity of tetracycline resistance determinants. *Plasmid* **3:**99–108.

McMurry, L. M., R. Petrucci, and S. B. Levy. 1980. Active efflux of tetracycline encoded by four genetically different tetracycline resistance determinants in *E. coli*. *Proc. Natl. Acad. Sci. USA* **77:**3974–3977.

Levy, S. B., B. Marshall, A. Onderdonk, and D. Rowse-Eagle. 1980. Survival of *E. coli* host-vector systems in the mammalian intestine. *Science* **209:**391–394.

Levy, S. B., G. B. FitzGerald, and A. B. Macone. 1976. Changes in intestinal flora of farm personnel after introduction of tetracycline-supplemented feed on a farm. *N. Engl. J. Med.* **295:**583–588.

Levy, S. B., and L. McMurry. 1974. Detection of an inducible membrane protein associated with R-factor-mediated tetracycline resistance. *Biochem. Biophys. Res. Comm.* **56:**1060–1068.

Uncloaking Hidden Viruses

In 1957, as a second-year Wesleyan University undergraduate student, I was riveted reading Raymond Latarjet's article in *Cancer Research.* He proposed that viruses could remain latent in human cells and cause cancer. This concept propelled me to examine the potential role of viruses in human diseases, particularly immune deficiency and cancer. I was at the time fascinated by the field of lysogeny, in which bacterial viruses (phages) remain hidden within the organism and can make products that are toxic to the host the bacteria invade (e.g., diphtheria toxin). Perhaps similar viral processes were involved in human disease.

Later, in medical school at Columbia University, I pursued this interest in lysogenic bacteriophages and was fortunate to do research in the Department of Microbiology chaired by Harry Rose. I worked with a newly hired professor, Herbert Rosencrantz, who had me evaluate how hydroxylamine activates infectious phage from a lysogenic state. He subsequently gave me a new compound, hydroxyurea, to look at the same process. It was during those studies that I was first introduced to the power of serendipity, when discovery is made by chance. In the presence of hydroxyurea, the bacteria I was culturing did not lyse (the phages were not activated), and there was no change in viable bacterial cell count. Nevertheless, the protein content of the culture increased tremendously. A simple observation under the microscope revealed very large bacteria in the treated culture that gave us the chance discovery that hydroxyurea blocked DNA, but not RNA, synthesis in these cells. Serendipity has since been a major theme in my research in microbiology.

During the summer of 1963, I continued pursuing my interest in lysogenic bacteria and worked with Elizabeth and Joe Bertani at the Karolinska Institute in Sweden. That experience showed me the exciting science going on at the Karolinska and enabled me to appreciate even further the types of interactions viruses can have with their host cell.

I felt poised to begin examining this type of interrelationship with mammalian cells, and the following summer I had a Louisiana State University fellowship that gave me this opportunity through studies of Burkitt's lymphoma in Africa. Reovirus type 3 had been isolated from Burkitt's lymphoma patients by Thomas Bell, and I went to the Entebbe East African Virus Research Institute in Uganda to investigate its possible etiologic role with Bell and Alexander Haddow.

It was in Entebbe that I learned how to grow mammalian cell cultures and to isolate mammalian viruses—quite a change from the bacterial cultures in New York. We took monkey kidneys and established our own cell cultures for recovery of viruses in the laboratory. During my four months in Africa, I traveled widely and witnessed the health conditions, visiting hospital wards and outside clinics. I examined children with Burkitt's lymphoma and patients with Kaposi's sarcoma. (This latter disease is now an important part of my AIDS research effort.) On my return to Columbia University for my final year at medical school, Ed Curnen, Chair of the Department of Pediatrics, offered me space in his virology laboratory to continue my Burkitt's lymphoma studies. That year, under the very able tutelage of Eru Tanabe, Curnen's diagnostic virologist, I learned many features of mammalian virology. Eru helped prepare me for this fascinating field of science. Our work that year suggested a connection between reovirus type 3 and Burkitt's lymphoma, but not as a causative agent.

After graduating from medical school, I went on to the University of Pennsylvania in Philadelphia, and during my internship and residency, I spent my spare time in research. At Children's Hospital with Gertrude and Werner Henle, I investigated the potential role of the newly described herpesvirus, Epstein-Barr virus (EBV), which is detected in Burkitt's lymphoma cells. At the Wistar Institute with Vittorio Defendi, I learned lymphoid cell biology. With the Henles, I was successful in showing a direct relationship of antibodies to EBV and the presence of Burkitt's lymphoma. Our surprise was the high prevalence anti-EBV antibodies in healthy children. This latter finding supported a common theme of the Henles that with all viral infections, there are many people who show no signs of the infection. At the Wistar, I helped derive human lymphoid B cell lines in the early days when only a few papers, including ours, reported the establishment of such cell lines. It was then fortuitous for me to be with the Henles when they showed the induction of prolonged growth of B cells by EBV. The irony came when I found EBV in the cell lines derived at the Wistar. Thus, in these two laboratories at different places in Philadelphia, my two research projects came together.

Further virus research training was obtained at the National Institutes of Health (NIH) from 1967 to 1970. This experience certainly laid the foundation for the major focus of my research career—retroviruses. Working with three giants in virology—Robert Huebner, Wallace Rowe, and Janet Hartley (I often called them "the triumvirate")—I learned important features of cell biology, virology, and virus pathogenesis. Initially, I drew from my experience with hydroxyurea and used the drug in attempts to increase production of human adenovirus proteins by infected cells. These viruses had just been reported by John Trentin to induce tumors in hamsters. Large quantities of viral proteins were needed to evaluate the possible role of adenovirus in human cancer. As predicted, with hydroxyurea, cell division was blocked and adenovirus protein synthesis was enhanced. Viral antigens thus became readily available for these studies.

I was soon introduced at NIH to the field of RNA tumor viruses (now called *retroviruses*) and took on the challenge of determining why murine sarcoma viruses (MSVs) showed two-hit kinetics of cell transformation in cultured mouse embryo cells but one-hit kinetics in rat cells. With Robertson Parkinson, I found that mouse embryo cells, once transformed by the MSV, could not grow into a visible focus of transformed cells without the spread of new progeny virus to other cells. MSV-infected mouse cells usually died. The spread of MSV, which was a replication-defective virus, depended on the replication competency of the "helper" murine leukemia virus (MLV) found in all MSV preparations. Thus, two particles were needed for MSV propagation. Since MSV-infected rat cells remained viable and could replicate the transformed state, MSV transformation in these cells showed one-hit kinetics.

This observation led me to question whether the virus resembling, by electron microscopy, an MLV in tissues of New Zealand Black (NZB) mice could have some biological function. These mice develop autoimmune disease and cancer, and a role of the virus was suspected. By cocultivating these NZB embryo cells with MSV-transformed hamster or rat cells, I was able to show that the NZB MLV could rescue the MSV and induce foci of transformation in rat cells but not in mouse cells. At first I believed that the absence of foci in mouse cells reflected an inability of the NZB virus to replicate and "help" in the spread of MSV in these cells. But soon I discovered, by serendipity, that this NZB MLV was not defective but represented an exception to the then current dogma of virology: that viruses infect cells from the animals from which they are recovered. The NZB MLV could not infect cells of its own host species (e.g., mouse) but could infect cells from a wide variety of other animal species such as rat and human.

To indicate this unusual tropism, I borrowed from the Greek and established the nomenclature of *xenotropism* (Gr. *xenos*, foreign; *tropos*, turning) for these viruses, *ecotropism* (Gr. *oikos*, environment) for viruses infectious for the cells of the host species, and *amphotropism* (Gr. *amphos*, both) for the dual-tropic type. The latter retroviruses, which could infect cells from many different animal species including mice, were isolated from wild mice by Rowe and Hartley and by Suraiya Rasheed and Murray Gardner.

On identifying the mouse xenotropic viruses, there was a "push" in the laboratory to isolate similar viruses from other mice. We readily recovered these viruses from tissues and embryos from all strains of house mice studied (*Mus musculus*), including wild house mice from San Francisco and elsewhere. (They were kept in cages under my desk in the office!) Why were these viruses present in mice since they could not spread and reinfect mouse cells? They could only be passed through the germ line, thus supporting the concept of an inherited virogene, but for what reason? The ultimate answer still is not known, but the subsequent finding that mouse serum neutralizes these viruses led some investigators to propose that neutralizing antibodies played a role in the virus infection.

Borrowing from my training in immunology, I could not understand how a virus inherited in the genes of a mouse and particularly expressed in its embryo would elicit a neutralizing antibody. Tolerance to the viral antigens, I thought, should be present. It was this curious

Stuart and Jay Levy

questioning that led to another serendipitous finding—that this neutralizing factor was not an antibody but a lipoprotein circulating in the blood of the mice. My laboratory then engaged in "fun" experiments moving the responsible antiviral apolipoprotein, discovered with the help of JoAnn Leong and John Kane, from mouse lipoproteins to human lipoproteins. The human lipoproteins could then neutralize the mouse xenotropic viruses. We searched for similar factors in human sera and for infectious human xenotropic viruses (not found), although human placentas showed evidence of retrovirus-like particles. This field of endogenous human virus research is again very active today.

It was an exciting time in the 1970s to have reported on these unusual xenotropic RNA viruses and their interaction with lipoproteins. I was suddenly invited to meetings of cardiovascular researchers. The finding of antiviral activity associated with lipoproteins offered support for some beneficial effect of the ingestion of certain fats, like olive oil. Soon we were engaged in research on the effect of nutrition on autoimmune disease and cancer and found that low-calorie and low-fat diets delayed the onset of NZB disease. It was important that, by these studies of viral and host factors, the concept emerged that inherited retroviruses might play a role in normal physiologic processes.

To directly evaluate the possible role of retroviruses in normal and abnormal development, I spent a sabbatical year (1977 to 1978) at the Weizmann Institute in Rehovot, Israel, and at the Pasteur Institute in Paris. At the Weizmann, I worked with Nechama Haran-Ghera on studies involving the role of murine ecotropic retroviruses in the causation of T cell leukemias. We studied how the virus could influence the evolution of the preleukemic cell into a leukemic cell. It was in this environment that I was further introduced into T lymphocyte biology; that experience laid the groundwork for my future work with CD4+ and CD8+ lymphocytes in HIV infection.

At the Pasteur Institute, I directly investigated how mouse retroviruses might influence normal developmental processes. Working with cultured murine embryonic carcinoma (EC) cells in the laboratory of François Jacob, I was able to show that these viruses could influence the differentiation of EC cells into nerve or bone cells. The virus studies were done in collaboration with Jean Claude Chermann in Luc Montagnier's unit. Thus, it was perhaps a quirk of fate that we current AIDS researchers were already working together in 1979 on retroviruses (murine type) at the time that the AIDS virus itself was emerging as a pathogen in Africa.

Returning to the United States, I continued my studies of retrovirus infection of EC cells. Then in 1981, AIDS hit San Francisco. My laboratory, which was very active in basic virologic approaches, was poised to take on the challenge of finding the cause. Cancer was our introduction to AIDS, and we attempted to culture Kaposi's sarcoma cells to find a responsible virus. Then, with the knowledge of immune deficiency in AIDS patients, we directed our efforts at looking for a causative virus in the blood.

My experiences in AIDS research over the past 18 years have convinced me that my past adventures in virology were meant to prepare me for this important challenge. All my previous work with animal viruses came to the forefront and helped me in studies that successfully isolated the AIDS virus, demonstrated its biological activities in cell culture and in animal species, and identified the role of the immune system (particularly CD8+ cells) in controlling the infection. Our early work with Kaposi's sarcoma has helped in our efforts to determine the role of human herpesvirus 8 in this cancer. And hydroxyurea is now a drug used against HIV.

Above all, my experiences growing up in virology were highlighted by certain key features that have characterized my career in medical research. High among these has been the benefit of very good mentors—leaders in the field who instructed me not only in the many facets of science but also, as role models, reaffirmed the importance of integrity and moral judgment. From the Henles, in addition, I learned valuable lessons in viral pathogenesis, particularly that there are always survivors of virus infections. These people can give insight into the pathologic process and potentially help those who are succumbing to disease. We now study HIV-infected people who have remained asymptomatic without therapy for over 20 years. An important secret of these long-term survivors is an effective CD8+ cell anti-HIV immune response. The work with the xenotropic viruses had me recognize differences in the cellular host range of viruses and had me question the original concept that HIV was a purely T lymphotropic virus. We showed the virus has a wide cellular host range including cells of the brain and the bowel. Importantly, my experience with EBV infection and the international travels, particularly in Africa, associated with my adventures in virology prepared me for the worldwide challenge of AIDS. I certainly learned in my early years of research to appreciate the need for global public health efforts to halt the spread of infectious diseases. The experience gained in directing efforts at controlling this new infectious disease epidemic should provide valuable insights into approaches for warding off future challenges of the microbial world.

JAY A. LEVY was born in Wilmington, Delaware, in 1938, the identical twin of Stuart B. Levy. He received his bachelor's degree from Wesleyan University in Middletown, Connecticut, in 1960. He later received a D.Sc. Honorary Degree from Wesleyan University in 1996. In 1961 he was a Research Fellow at the Université de Paris. He received his medical degree from Columbia University in 1965. He did his residency at the Hospital of the University of Pennsylvania. His first professional position was as Staff Associate at the National Cancer Institute from 1967 to 1970. In 1970, he was part of the team that searched for Epstein-Barr virus and infectious hepatitis virus in wild chimpanzees in the forests of Uganda. He held a NATO Fellowship in Paris and was a visiting scientist at the Weizmann Institute of Science and the Institut Pasteur. Since 1985, he has been Professor in the Department of Medicine and Research Associate in the Cancer Research Institute at the University of California, San Francisco (UCSF). He is the Director of the Laboratory of Tumor and AIDS Virus Research at UCSF. He has received a variety of awards, including Fulbright and French Government Awards in 1960 and 1961, the Murray Thelin

Award of the National Hemophilia Foundation in 1986, and the Award of Distinction from the American Foundation for AIDS Research in 1994. He is a Fellow in the American Association for the Advancement of Science (1993) and the Infectious Diseases Society of America (1996). In 1998, the *San Francisco Chronicle/Examiner* named him one of the ten most influential people in the Bay Area.

The following books and papers are representative of his more recent publications:

Levy, J. A. 1999. Xenotropism: The elusive viral receptor finally uncovered. *Proc Natl. Acad. Sci. USA* **96:**927–932.

Stranford, S. A., J. Skurnick, D. Louria, D. Osmond, S. Y. Chang, J. Sninsky, G. Ferrari, K. Weinhold, C. Lindquist, and J. A. Levy. 1999. Lack of infection in HIV-exposed individuals is associated with a strong CD8(+) cell noncytotoxic anti-HIV response. *Proc. Natl. Acad. Sci. USA* **96:**1030–1035.

Barker, E., C. Mackewicz, G. Reyes-Terán, A. Sato, S. Stranford, S. H. Fujimura, C. Christopherson, S. Y. Chang, and J. A. Levy. 1998. Virological and immunological features of long-term human immunodeficiency virus-infected individuals who have remained asymptomatic compared with those who have progressed to Acquired Immunodeficiency Syndrome. *Blood* **92:**3105–3114.

Barker, E., K. N. Bossart, and J. A. Levy. 1998. Primary CD8+ cells from HIV-infected individuals can suppress productive infection of macrophages independent of beta-chemokines. *Proc. Natl. Acad. Sci. USA* **95:**1725–1729.

Levy, J. A. 1998. *HIV and the Pathogenesis of AIDS,* 2nd ed. ASM Press, Washington, D.C.

Levy, J. A. 1997. Three new human herpesviruses (HHV6, 7, and 8). *Lancet* **349:**558–563.

Mackewicz, C. E., E. Barker, G. Greco, G. Reyes-Teran, and J. A. Levy. 1997. Do β-chemokines have clinical relevance in HIV infection? *J. Clin. Invest.* **100:**921–930.

Levy, J. A. 1996. Infection by human immunodeficiency virus—CD4 is not enough. *N. Engl. J. Med.* **335:**1528–1530.

Lennette, E. T., D. J. Blackbourn, and J. A. Levy. 1996. Antibodies to human herpesvirus type 8 in the general population and in Kaposi's sarcoma patients. *Lancet* **348:**858–861.

Blackbourn, D. J., C. E. Mackewicz, E. Barker, T. K. Hunt, B. Herndier, A. T. Haase, and J. A. Levy. 1996. Suppression of HIV replication by lymphoid tissue CD8+ cells correlates with the clinical state of HIV-infected individuals. *Proc. Natl. Acad. Sci. USA* **93:**13125–13130.

Barker, E., C. E. Mackewicz, and J. A. Levy. 1995. Effects of TH1 and TH2 cytokines on CD8+ cell response against human immunodeficiency virus: Implications for long-term survival. *Proc. Natl. Acad. Sci. USA* **92:**11135–11139.

Levy, J. A. 1992–1995. *The Retroviridae*, vols. 1–4. Plenum, New York, N.Y.

Barnett, S. W., K. K. Murthy, B. G. Herndier, and J. A. Levy. 1994. An AIDS-like condition induced in baboons by HIV-2. *Science* **266:**642–646.

Experiences of a Virus Hunter

For the last 30 years I have been chasing outbreaks of disease—viral diseases that are as rare and as deadly as any that have ever plagued humankind. And I have been told over and over that virus hunters like myself soon would no longer be needed; like the dinosaurs we would soon disappear along with the viral diseases that have preoccupied me. But the viruses do not seem to have gotten this message. So as my colleagues and I age and gray, new viruses emerge: Ebola, Nipah, hantaviruses, West Nile, HIV, and others—all viruses that wreak havoc on societies, cause individual human suffering, and evoke the age-old fears of epidemic plagues—all challenges and always changing.

It has been a long way from the dry plains of West Texas where I was raised to where I am now in Atlanta, Georgia, at the Centers for Disease Control and Prevention (CDC). How do you recall this past from the 1950s for a student when everything seems so different today? Was it a different naive time, my youth, or something about myself? In any case, I came to science from circumstances rather than any considered decision. I never consciously thought about role models or careers during high school, but in a lower middle-class boyhood in West Texas, the oil fields dominated. If there were male English teachers in my school, I don't recall; there were no poets; and there weren't even serial killers in the newspaper stories. However, I had been curious and an avid reader as long as I can recall—it didn't matter whether it was fiction, fantasy, travel, anthropology, biology, or almost anything on the shelves of the library.

Even before I knew I wanted to study science, my nature led me to look for empirical answers to life's mysteries. How could you prove what was right or wrong or be sure that God really existed? These questions continued to bother me, and I kept wondering when and where I could get to the truth. But what I have found over the years is that science is as much a belief system as any religion, and each belief system is best left in its own domain, rather than trying to usurp the ground of the other. Science is about ordering and understanding our world; it's not about existential truths.

Several high school teachers stand out in my memory, but one who made the greatest impact was in biology. She was a demanding person in class and had a very straightforward way of telling you where you stood—an unsettling experience for a high school sophomore. My hands sweated under the spotlight of her exams until I dissolved the black coating on the desks cum dissection tables and I turned in wrinkled, stained test papers. One day I asked her, "How do you type blood?" I can't remember why I was so interested or what gave me the courage to approach her. But my interest pleased her because she already had arranged for an after-school apprenticeship for a couple of her students with the clinical lab at the hospital.

I went there at her urging and met another person who would profoundly influence my future, Zorus Colglazier. His unlikely name may have contributed to his flamboyant personality. He had been introduced to the medical field in the Navy and afterward studied at a private school of medical technology. When I arrived in the hospital lab one evening, he was dressed in old white bell-bottom trousers and a T-shirt, covering the 3 to 11 shift. I am not sure what he saw in me or why he agreed to have me as a hanger-on, but I soon was a regular after school or in the evenings. We rarely discussed personal issues, but as I look back, I can see the situation in my mind's eye. Zorus was too bright and too ambitious to stay in the hospital lab environment and was moving along to study part-time at the local junior college, escape from an unhappy marriage, and eventually become a physician.

I was lucky to be a channel for some of his intellectual energies and talents as a teacher. He was no-nonsense and had a sarcastic edge reserved for stupid actions. I responded with the total enthusiasm that came from the intellectual challenge of learning the laboratory tests and the responsibility of working in an exciting environment where your work was not graded with a number but simply had to be correct for the patients' sake. I worked my way through the sequential apprenticeship of stool exams, urine exams, and finally blood counts and chemistries—a hierarchy graded mainly on smell. It was heady stuff for a high school student, and I became thoroughly hooked on the mental challenges, the drama, the human interplay. I am sure this couldn't be done today—just think of insurance, liability concerns, and general "CYA" (a phrase I learned later in my career and that is, of course, increasingly useful in describing the world in the year 2000).

As I struggled through adolescence and looked around for a college to attend, I had already formed a bond with Rice University in Houston. Sputnik had led to a summer course in math offered in that particular tropical swamp, and another high school teacher had nudged me in that direction. I was so taken with the ideas and the atmosphere that I didn't even notice the heat or the humidity. Still, it was not surprising that I gravitated into a medical school track after I entered Rice. I subscribed to the *New England Journal of Medicine* and eagerly slogged my way through many of the articles. I stumbled upon Zinnser's *Rats, Lice, and History* in the library and haunted Majors' medical bookstore in downtown Houston, where I picked up Chauncy Leakes' reprint of Ashbel Smith's work on the 1839 yellow fever outbreak in Galveston—something to appeal to my budding medical interests and to the inevitable chauvinism of all Texans. Rice kept me from making a common mistake—there was no premedical major. I was a chemistry major and pursued that as a discipline rather than electing the courses that were shadows of the future medical school curriculum. I think mastery of a field before graduate education is important to the aspiring physician or microbiologist. In my case, the most valuable course for really understanding the scientific basis of medicine was a thorough grounding in physical chemistry and thermodynamics.

In my senior year, I consulted my faculty advisor in the chemistry department and told him I was going to apply to medical school. He looked up from his desk and said, "If you decide to go into a real profession, come back and see me." I thanked him for his time and went to the library to get the addresses to write for medical school catalogs. I had a couple of criteria. Another Sputnik spinoff had sent me to a brief summer course in Boston organized by MIT and Harvard that centered around biophysics, so I was enthusiastic about moving further east to get to know that part of the United States. I also wanted a medical school with a research orientation and an experimental slant toward education. Johns Hopkins University seemed to fill the bill.

Mine is probably the last generation for whom medicine was an art and the first for whom it approached a science. Osler's philosophy permeated the teaching at Johns Hopkins: "He does not see the pneumonia case in the amphitheater from the benches, but he follows it day by day, hour by hour; he has his time so arranged that he can follow it; he sees and studies similar cases and the disease itself becomes his chief teacher, and he knows its phases and variations as depicted in the living; he learns under skilled direction when to act and when to refrain." This is still a valuable lesson for anyone entering the public health field faced with the outbreak of an unknown disease among a patient base you know little about.

Osler also preached the Hopkins triad of patient care, research, and teaching, but I'm not sure it works today. I couldn't juggle the combined demands of a full-fledged clinical career in today's academic practice meat grinder and of directing a well-funded research program in, for example, molecular biology. One of today's challenges is to solve the interfaces and integrations of the multiple disciplines that are needed to attack the most challenging problems.

I made another lucky choice: I returned to Texas to do internal medicine training at Parkland Hospital. I went because I had done summer work there and found the young, dynamic faculty to my liking. I have since seen the young faculty there reincarnated as the gray-bearded authors of many of the standard textbooks on the shelves today. In addition to the intellectual ferment, there was a "just-the-facts-Ma'am" attitude and a sharp logical approach that was much to my liking. Parkland was an excellent complement to the Hopkins experience in many ways. But the most important thing that I acquired from Parkland was my acquaintance with Jay Sanford, who exemplified many of the best things there. I met him off and on through the rest of my career, but he had a much greater influence than total "face time" might account for. He was among several people who dwelled in my psyche and who were available to ask, "What would Jay (or *x*) do in this fix?"

I was accepted into the U.S. Public Health Service (PHS) during my senior year at Hopkins, with the idea that I would begin after two years of internal medicine training. The PHS had several programs, and I had trouble choosing which of two to apply for: one headed by Bob Chanock that involved studying respiratory and enteric viral diseases as they broke out in schools and communities, the other a position working with Karl Johnson at the National Institute of Health's Middle America Research Unit (MARU) in the Panama Canal Zone studying Bolivian hemorrhagic fever and other tropical viruses. I saw this as a choice in which neither was obviously better, neither would close any doors, and I could learn a lot from both. I figured, I've been through four high-pressure years of college, four years of higher-pressure medical school, and two years of internship and residency where I had almost no time to do anything, so I decided to go to Panama, join the PHS, and see the world!

The lab at MARU handled some of the most infectious viruses known to medicine, long before sophisticated equipment, protocols, and procedures had been developed to handle such biohazardous agents. Not long after my arrival, a new opportunity for field research presented itself in the form of an outbreak of Venezuelan equine encephalitis (VEE), which was decimating horses and burros in Central America. Discovered in the 1940s, VEE breaks out in abrupt, aggressive epidemics that are lethal to equines. Then it disappears, only to return some unpredictable number of years later. Mosquitoes spread VEE—and bite humans as well as horses—so humans are also susceptible to the disease. We thought if we had a chance to make a difference in containing the epidemic, we absolutely should try to do so. The U.S. Army had developed a vaccine to protect our troops should this ever be used as a biological warfare agent, and Karl Johnson believed that by immunizing the horses we could prevent the spread of the virus to humans.

So my first full-fledged field assignment as your basic foot soldier began. I have to confess that my outward professional demeanor did not match the "little kid" excitement that I felt. Our mission was to study the disease in horses and in people. To bleed horses, we had to bribe Costa Rican cowboys, unless, of course, one was a lot more skillful at catching wild horses than the average PHS doc. It was odd to see at first, but later I'd see this a lot, particularly in African villages: People and animals are bled in the order of importance. In a village where they think the bleeding will be good for them, they'll let you bleed the men, the boys, the adult women, and then the girls. In places where they're afraid something bad will happen, you'll bleed the girls, the adult women, the boys, and then—if everything has gone well up to that point—maybe you'll get the adult men.

Today, now that a lack of funding has closed MARU and other labs like it are struggling, I wonder who will be in these places to spot the threats of the future. Who has the access and the experience to deal with these diseases before they reach our country? VEE proved to me that we can deal with these diseases now, where they come from, or (as was almost done with VEE) we can wait until we see them in our own backyard and start from square one after our own citizens start getting sick and dying. And that's a crying shame.

One day Karl Johnson came in and set two vials of reagents on my desk and said, "I think this is real; see what you can find out about Australian antigen in Panama." I was to study three Indian groups in South America, so I gathered the scientific equipment I'd need, making sure to pack my swimsuit and snorkeling gear, too. While I was involved in the study of hepatitis, a mysterious epidemic, possibly a viral hemorrhagic fever, hit Cochabamba, Bolivia. A nurse had been exposed, and later a pathologist cut himself during her autopsy and also developed the disease, which eventually killed him. The hospital where I was caring for him before his death was very tense. I always joke that this is one business where you can usually get a table to yourself in the cafeteria wherever you are, and this was certainly the case in Cochabamba. The experience left a lasting interest in the hemorrhagic fevers. These were viruses that could take anyone down; they were poorly understood so they furnished a real challenge, and understanding their epidemiology was both urgent and a fascinating ecological problem.

The Scripps Clinic and Research Foundation in La Jolla had a wonderful research program in viral immunology. Having seen firsthand how field epidemiology fit into the study of viral diseases, I was ready to learn how the laboratory brought the picture into sharper focus. Three years at a crossroads of immunology didn't turn me into a real immunologist but did give me the tools to understand the problems of my favorite viruses as they interacted with their natural hosts and with the unfortunate humans who got in their way. If I had any lesson from my career for young persons, it would be to associate themselves with people of quality doing significant work in several fields, as I had been fortunate to do so far in my career. Don't close any doors if you don't have to, but always try to contribute and not just be a hanger-on.

At that point in my career, I was in danger of being seen as the world's oldest living dilettante: I'd done internal medicine, then some epidemiology and tropical virology, some more internal medicine, and immunology. I was rapidly approaching the point where I couldn't afford to dabble in another specialty and still hope to be taken seriously when an old Army friend from Panama made me an offer I couldn't refuse—join the Army and go to work at the United States Army Medical Research Institute of Infectious Diseases (USAMRIID) at Fort Detrick, Maryland. "Your own Biosafety Level 3 laboratory. You don't have to worry about anyone else working in the lab. You'll have a young G.I. as a technician, as well as a captain who just finished his Ph.D. in biochemistry. And I'll also throw in all the fetal calf serum you can use." So I officially entered the U.S. Army as a major, getting credit for my years of uniformed service in the PHS.

Not long after I finished what passed for basic training for medical corps officers, there was an outbreak of Rift Valley fever (RVF) in Egypt. This was a virus active in sub-Saharan Africa but not known to be present anywhere else, including Egypt. Naval Medical Research Unit No. 3 in Cairo was working with isolates from the epidemic. They needed to protect their people, including Americans and Egyptians. We needed to get enough vaccine over there to immunize them. It could also be an opportunity to field-test the vaccine and perhaps run a trial in communities in Egypt where the disease was being transmitted. I balanced the pros and cons of going to Egypt. Probably a lot of sane, rational people would find this a no-brainer: head into a Third World epidemic where people are dying? No, thank you. But I figured we had a vaccine that should protect me, so danger from the virus should be minimal. This was a fascinating disease in a part of the world I'd never been to. There was an opportunity for scientific progress if we could use this vaccine to protect people and figure out how the virus got there and why it was causing hemorrhagic fever. I agreed to go.

Once I arrived, I realized that I hadn't considered that I was an Army research scientist working in a Navy lab. Naively, I thought of all of us as military. The Navy had let me come because they needed vaccine to protect the lab crew from a deadly disease, but they weren't particularly happy to have some Army interloper poking around in their territory. The head of the lab still gave me free rein, and everyone was vaccinated. The Egyptian government was unimpressed with an experimental vaccine and would neither cooperate nor authorize us to study the disease. Still, I learned a lot, got to work with some Egyptian and some U.S. Navy scientists who are still friends today, and maybe even helped understand a little more about RVF.

After I returned to USAMRIID, I dug into the experimental aspects of RVF. Cedric Mims had done the last meaningful experiments on pathogenesis in the 1950s, so there was a chance to apply some new techniques and ideas to the disease. In fact, we found an antiviral drug that worked in vitro in laboratory animals. By 1980, we knew ribavirin had potential for treating patients with RVF, but to date we have still not had a good chance to test it in humans. For one thing, it's hard to predict when the next epidemic of Rift will occur, and you'd need an epidemic to get enough patients for an effective study, because only 1 or 2% of the people who get RVF develop full-blown hemorrhagic fever. And as with most epidemics in faraway regions, if you wait for one to happen, it's usually peaked by the time you get over there.

The final piece of the mystery was: where did it come from? Where was the virus hiding between epidemics? It turns out that the maintenance of RVF in sub-Saharan Africa was related to a combination of weather conditions and the habits of several species of so-called floodwater *Aedes* mosquitoes. After the virus was rescued from the dried eggs by flooding, it could spread in typical arbovirus mosquito-host-mosquito-host fashion. Floodwater *Aedes* were the critical reservoirs, although a variety of mosquitoes could be vectors once they fed on a viremic animal. This is a vivid example of how nature can replicate the same patterns using a different but equally specific set of viruses and host organisms around the world; studies of Keystone and La Crosse viruses in the United States had laid the foundation for the RVF studies. It also speaks to how much we can learn through surveillance of viruses and study of disease patterns. This hypothesis explains why the virus died out in Egypt after the epidemic. After the Aswân Dam was constructed, there was no more alluvial flooding. There's also virtually no rain in Egypt, so there are very few floodwater mosquitoes. As sheep and cows died out or became immune in their recovery, it became impossible for the virus to survive simply by jumping from host to host. Without its specific floodwater mosquitoes, the virus can't maintain itself over the long haul. We had found out why the outbreaks occur where they do and we can now even predict them to some degree through satellite remote sensing. We have a vaccine proven to protect against the virus, and we've begun setting things up so that when it hits again, we have an experimental drug ready to be tested in the most severely ill patients. The answers don't always come quickly, and they almost never come cheaply, but if they translate into lives saved and countless people spared suffering, then I'd say it's well worth the effort.

In 1985 I made several forays into central Africa as part of a team looking for Marburg, Ebola, and other hemorrhagic fever viruses. We needed to find out the sources of these viruses and find out why the outbreaks suddenly appeared and just as suddenly disappeared. We sought out physical clues, taking blood samples wherever we could, looking for antibodies that might lead us to the sources of exposure. This sort of effort presents problems for medical ethics. Although our research had long-term potential benefits for inhabitants of central Africa, it certainly had no immediate benefit for any of the individuals we needed to study. We wanted blood and could not promise any reward. Cooperation depended on our showing a real sense of caring.

Although we still hadn't figured out the reservoir for Ebola virus in the Congo, we soon faced another problem much closer to home. A shipment of monkeys in Reston, Virginia, began dying, and to our alarm we found a filovirus—one that initially appeared to be the lethal African Ebola virus. Had our worst nightmare come true? Was a deadly hemorrhagic fever about to devastate the United States? Fortunately it proved to be a less pathogenic sub-type, but the threat remains. Deadly viruses can move around the world. This was one of my last investigations at USAMRIID—my career was about to march on.

But before I left USAMRIID, I was given an honor that meant a tremendous amount to me: they created an award in my name. Underneath the USAMRIID insignia with its motto, "Research for the Soldier," the plaque reads:

> In the inception and granting of this award, we honor Colonel C. J. Peters, Medical Corps, a brilliant and innovative scientist, scholar, and physician, and an example of unfailing dedication to the protection of the health and well-being of our Armed Forces. The recipient of this award follows in the tradition of Colonel Peters' persistence and creativity in the search for pre-

> vention and therapy of militarily relevant diseases, and reflects his enthusiasm for surmounting the obstacles inherent in the struggle to advance medical knowledge. Most important, the recipient of this award is a mentor, inspiring younger researchers to participate and succeed in the challenge and to share that profound reverence for life that is illustrated by the career of Colonel Peters, whose work is a lesson not only in science, but in humanity.

I was moved and humbled. If any fraction of what it said about me was true, then maybe I had accomplished some of what I set out to do.

And then it was 1991 and time to move on to a position at the CDC and the chance to work in the public health arena. It didn't take long for an opportunity to present itself. In 1993 Native Americans suddenly began to die in the southwestern United States in the Four Corners region. Sudden deaths caused by acute respiratory distress syndrome in otherwise healthy individuals raise alarm bells. And the pattern of this disease was like none other before it. Bacteria, viruses, chemical poisons—at that time it was impossible to rule out any cause. Some of the laboratory findings from the field were consistent with an arenavirus infection. The patients had low blood platelet counts and shock, both of which fit the pattern. But what spoke against these particular viruses was the tremendously large number of white cells—part of the body's defense mechanism—circulating in the blood, in some cases as high as 50,000. I knew of one type of virus that regularly caused the hemorrhagic fever syndrome and high white counts: hantaviruses. This was a new hantavirus—one that the Special Pathogens Branch and a large number of collaborators identified as the cause of the disease. So in January 1994 we registered its name: Muerto Canyon virus—the virus of the canyon of death.

In a stunningly short time, all hell broke loose, and I found myself embroiled in a political morass. Representatives of the National Park Service were upset because they already had a location called Canyon de los Muertos—Canyon of the Dead Ones—and felt an association with a deadly virus could possibly be bad for tourism. Native American groups didn't like it because the Canyon de los Muertos had been the site of an Indian massacre and they didn't want the enormity of that act overshadowed by any other association. Ultimately, after going back and forth on a wide range of possibilities, we decided on the one name that nobody was completely happy about but everyone could live with. The Navajo Tribal Council approved it unanimously. The designation? Sin Nombre virus: No Name virus. For the killer without a name.

C. J. PETERS was born Clarence James Peters, Jr., on September 23, 1940, in Midland, Texas. He chose to abandon the formal name Clarence and nicknames like "little Pete" and to be known to all as C. J. He received a bachelor's degree from Rice University in 1962 and a medical degree from Johns Hopkins University School of Medicine in 1966. He went on to a career in the government, working for the U.S. Public Health Service, the U.S. Army, and most recently the Centers for Disease Control and Prevention. He is truly a pioneer physician and virus hunter, having ventured into remote regions of the world in search of the most deadly hemorrhagic fever viruses. His is a career of adventurous epidemiology and individual caring as a physician. He currently is Chief of the Special Pathogens branch of the CDC.

The following books and papers are representative of his publications:

Linthicum, K. J., A. Anyamba, C. J. Tucker, P. W. Kelley, M. F. Myers, and C. J. Peters. 1999. Climate and satellite indicators to forecast Rift Valley fever epidemics in Kenya. *Science* **285:**397–400.

Ksiazek, T. G., P. E. Rollin, A. J. Williams, D. S. Bressler, M. L. Martin, R. Swanepoel, F. J. Burt, P. A. Leman, A. S. Khan, A. K. Rowe, R. Mukunu, A. Sanchez, and C. J. Peters. 1999. Clinical virology of Ebola hemorrhagic fever (EHF): virus, virus antigen, and IgG and IgM antibody findings among EHF patients in Kikwit, Democratic Republic of the Congo, 1995. *J. Infect. Dis.* **179:**S177–187.

Peters, C. J., and J. W. LeDuc. 1999. An introduction to Ebola: The virus and the disease. *J. Infect. Dis.* **179:**9–16.

Peters, C. J. 1998. *Virus Hunter: Thirty Years of Battling Hot Viruses Around the World.* Anchor Books, New York, N.Y.

Sanchez, A., S. G. Trappier, B. W. Mahy, C. J. Peters, and S. T. Nichol. 1996. The virion glycoproteins of Ebola viruses are encoded in two reading frames and are expressed through transcriptional editing. *Proc. Natl. Acad. Sci. USA* **93:**3602–3607.

Peters, C. J. 1997. Pathogenesis of viral hemorrhagic fevers, pp. 779–799. *In* N. Nathanson, R. Ahmed, F. Gonzalez-Scarano, D. Griffin, K. V. Holmes, F. A. Murphy, and H. L. Robinson (ed.), *Viral Pathogenesis*. Lippincott-Raven Publishers, Philadelphia, Penn.

Duchin, J. S., F. T. Koster, C. J. Peters, G. L. Simpson, B. Tempest, S. R. Zaki, T. G. Ksiazek, P. E. Rollin, S. Nichol, E. T. Umland, R. L. Moolenaar, S. E. Reef, K. B. Nolte, M. M. Gallaher, J. C. Butler, R. F. Breiman, and the Hantavirus Study Group. 1994. Hantavirus pulmonary syndrome: a clinical description of 17 patients with a newly recognized disease. *N. Engl. J. Med.* **330:**949–955.

Nichol, S. T., C. F. Spiropoulou, S. Morzunov, P. E. Rollin, T. G. Ksiazek, H. Feldmann, A. Sanchez, J. Childs, S. Zaki, and C. J. Peters. 1993. Genetic identification of a hantavirus associated with an outbreak of acute respiratory illness. *Science* **262:**914–917.

Peters, C. J., E. D. Johnson, P. B. Jahrling, T. G. Ksiazek, P. E. Rollin, J. White, W. Hall, R. Trotter, and N. Jaax. 1991. Filoviruses, pp. 159–175. *In* S. Morse (ed.), *Emerging Viruses*. Oxford University Press, New York, N.Y.

Peters, C. J., P. B. Jahrling, C. T. Liu, R. H. Kenyon, K. T. McKee, Jr., and J. G. Barrera Oro. 1987. Experimental studies of arenaviral hemorrhagic fevers, pp. 5–68. *In* M. B. Oldstone (ed.), *Current Topics in Microbiology and Immunology*, vol. 134. *Arenaviruses: Epidemiology and Immunotherapy*. Springer-Verlag, Heidelberg, Germany.

Peters, C. J., J. A. Reynolds, T. W. Slone, D. E. Jones, and E. L. Stephen. 1986. Prophylaxis of Rift Valley fever with antiviral drugs, immune serum, an interferon inducer, and a macrophage activator. *Antiviral Res.* **6:**285–297.

Peters, C. J., and A. N. Theofilopoulos. 1977. Antibody dependent cellular cytotoxicity against murine leukemia viral antigens: studies with human lymphoblastoid cell lines and human peripheral lymphocytes as effector cells comparing rabbit, goat, and mouse antisera. *J. Immunol.* **119:**1089–1096.

Peters, C. J., R. W. Kuehne, R. Mercado, R. H. Le Bow, R. O. Spertzel, and P. A. Webb. 1974. Hemorrhagic fever in Cochabamba, Bolivia, 1971. *Am. J. Epidemiol.* **99:**425–433.

Peters, C. J., W. C. Reeves, V. Rivera, and K. M. Johnson. 1973. Epidemiology of hepatitis B antigen in Panama. *Am. J. Epidemiol.* **98:**301–310.

The Path of Biomedical Research

Although I enjoyed both clinical medicine and biomedical research, I soon realized that I had to make a choice if I was to perform at a level of excellence. I wanted to make use of both my doctoral and medical degrees by pursuing a career in biomedical sciences. Nevertheless, you can't do everything—at least not if you want to do things well. I selected the research path—to discover the knowledge on which medicine is practiced. I chose to focus on human viruses, which has allowed me to combine my interests in virology and human health. My research on viruses began with bacterial viruses while I was studying at Johns Hopkins University. When I went on to work at the National Institutes of Health (NIH), I focused on animal viruses and was the first to isolate a polymerase encoded by an animal virus.

In considering the establishment of my career, I cannot overemphasize the role of mentoring. Several individuals played key roles as mentors in my development as a microbiologist. As an undergraduate at Harvard, I fortunately worked in Paul Doty's laboratory with Julius Marmur, then a senior fellow there. My tutor, Jacques Fresco, also in the Doty lab, led me to that opportunity. James Watson taught me about DNA structure and was still relatively junior. Francis Crick was at Harvard on sabbatical. The big research question was to find the messenger (messenger RNA was found several years later). In that early time, it was actually possible to read all the papers published that mentioned DNA.

From Harvard, I went to Johns Hopkins for medical school in a special program directed by Barry Wood, one of the great microbiologists of the time (and Harvard's last All-American quarterback). Barry took care of me as a faculty member when I returned to Hopkins after the NIH. Possibly most critical was my Ph.D. mentor during this time, Charles A. Thomas, Jr., who did groundbreaking work on the structure of bacteriophage genomes. I spent my first year at Hopkins working almost full time in Charlie's lab, and it was so stimulating that I took leave after the first year of the regular medical curriculum to earn a Ph.D. At the NIH, Arthur Weissbach and Norman Salzman gave me great support. At a later stage in my career, Robert Marston, a previous NIH director who was then President of the University of Florida; Harlyn Halvorson, then Chair of the Public and Scientific Affairs Board of the American Society for Microbiology; and Robert Petersdorf, the President of the Association of American Medical Colleges, played a similar role for my involvement in public affairs. The pleasure all of these men took in the development of the next generation had a powerful effect on my desire to do the same.

My first faculty position was in the Microbiology Department at Hopkins with Barry Wood as the chair. Three years later, he died prematurely and was succeeded by Dan Nathans. I was full of suggestions for things that Dan should do as chair. He always listened and even did as I suggested on several occasions. Thus, when I was offered the chair at the University of Florida College of Medicine, I thought that either I should accept and take the responsibility for all of my "bright" ideas or keep quiet. Since the latter was unlikely, I moved to Florida. I served as chair at both Florida and Cornell University Medical College for a total of twenty-one years. The position was highly satisfactory at both institutions. The chance to create academic environments where research and teaching can flourish, where faculty develop and promote student interest, and where students become scientists has been extremely gratifying.

My research for the past twenty-eight years has focused on the molecular mechanisms underlying the replication of a cryptic human virus, adeno-associated virus (AAV). The direction the work has taken in recent years is a good illustration of the fact that it is impossible to predict the practical outcome of fundamental research. Although all of us are infected by AAV, the virus has never been associated with any human disease and thus was little known until recently. It was not discovered until 1965, when it was found as a contaminant of supposedly pure preparations of adenovirus, a common human pathogen. It was quickly shown that AAV required coinfection with an adenovirus for productive infection in cell culture. In 1968, while I was in the U.S. Public Health Service at NIH, I met Jim Rose at a party, and the next day he presented me with a problem. He had recently characterized the purified AAV genome as a linear duplex molecule of 3×10^6. Yet Lionel Crawford had just published a study showing that the genome in the virus particle was a linear single strand of half that molecular weight. Crawford suggested that an unlikely possibility for resolution of this apparent paradox was that half the virions contained plus DNA strands and the remainder contained minus strands. On purification the complementary strands would anneal to form duplex DNA. Jim and I quickly showed that this unlikely possibility was indeed true. We mixed virus particles containing normal-density DNA with particles containing heavy-density DNA. When the DNA was purified, the double-stranded DNA we isolated had a hybrid density.

From that fundamental beginning my research involved the details of the viral DNA structure, including the determination of the nucleotide sequence in 1983 and studies on DNA replication and the regulation of gene expression. However, much of the effort has been devoted to the molecular characterization of the establishment of latent infection by AAV. Originally, AAV was thought to be defective because it required a helper virus coinfection in cell culture. We now think that the helper requirement is more a reflection of the fact that the virus life cycle involves integration of the viral DNA in the host cell genome. If the host cell is healthy, the virus stays in the latent state. However, if the cell is severely stressed, the integrated viral genome is activated (i.e., gene expression occurs), rescued, and progeny virus made. Initially we discovered that integration had occurred. Only two mammalian viruses integrate as a normal part of the life cycle: retroviruses and AAV. As we showed in 1990, AAV is the only virus that integrates at a specific site in the human genome. Most recently we have been able to refine the required site for integration to a 30-nucleotide sequence and have shown two signal sequences within this stretch that must be intact.

Because it does not cause disease, AAV was not initially of great medical interest; this situation changed with the recognition that the virus could integrate its genome. Coupled with the virus's persistence, this characteristic made it a strong candidate to serve as a vector for human gene therapy. Now clinicians and entrepreneurs in large numbers are interested. As a consequence, work in my laboratory has been supported by industry, as well as by NIH, and I serve as a consultant. At the scientific level my work on AAV has been seminal in the current interest in using viruses as a vector for human gene therapy. Research in my laboratory has played a major role in the understanding of the molecular biology of both productive and latent infection. At the public level I have been very involved in the early discussion about the safety of recombinant DNA technology and continue to be so.

Besides my activities as a biomedical research scientist and educator, I have been highly involved in public affairs and served as Chair of the Public Affairs Board of the American Society for Microbiology (ASM) from 1989 to 1995. My early experiences made me sensitive to events at the national level that affected the ability to do research in microbiology and science in general and also were related to the consequences of science to the public. For this reason I have been active on the national and even international levels in various microbiological societies and organizations involved in medical education. In the same vein I am pleased to have served in numerous advisory capacities for the government. I strongly believe that participation by scientists in public affairs is critical for the well-being of both the scientific enterprise and the public. At the public level it is difficult to take individual credit for what must be a community effort at all times. I can claim to have been an active force in furthering biomedical research and ensuring the safe development of recombinant DNA technology. For the past fifteen years I have been especially devoted to ensuring the adequacy of funding of biomedical research at NIH and at educational institutions. I consider the funding of biomedical research and of microbiology to be a national responsibility. When the Reagan administration withheld funds that had been appropriated by Congress, I fought hard to see that funding restored. A lawsuit was prepared that challenged the President's decision to withhold funds from NIH. On the day before the suit was to be filed in federal court, the Reagan administration relented and released the fund to NIH. Some blame me for initiating the suit.

As a consequence of my interest in public affairs and medical education, I represented both the Association of Medical School Microbiology and Immunology Chairs and the ASM in the Council of Academic Societies (CAS), one of the three major components of the Association of American Medical Colleges. Eventually I became Chair of the CAS, which led to my being selected to chair the overall AAMC. In this role I was immersed in all of the problems facing medical schools, from support for basic research to the impact of managed care. Thus when I was offered the opportunity to return to Florida as the Dean of the College of Medicine in 1997, I viewed it as an interesting challenge, as well as a chance to reunite with many old friends. The real question was whether the College of Medicine could continue to function as a true academic institution in the face of current economic pressures. So far we have been able to maintain an environment conducive to scholarly endeavor and to excellence in patient care. In 1999 I also served as the Interim Vice President for Health Affairs, which deeply involved me in the health care side of the equation. This has been satisfying in the sense that it completes a spectrum of experiences from basic research to health care delivery. As I suggested earlier, indeed one never knows where basic research may lead.

Clearly I have led an active and diverse career in microbiology in which I have combined my interests in research, medicine, virology, education, and public service. Do I wish that I had made different choices at junctures of my career? No!

KENNETH BERNS was born in Cleveland, Ohio, in 1938. He was educated at Johns Hopkins University and has doctoral and medical degrees. He has worked at the National Institutes of Health and as a Professor at the College of Medicine at the University of Florida. He was President of the American Society for Microbiology from 1995 to 1996 and President of the American Society for Virology from 1988 to 1989. He is a member of the National Academy of Sciences. He served as Professor and Chair of the Department of Microbiology of the Medical College of Cornell University in New York. He is also a member of the Institute of Medicine. He currently is Dean of Medicine and Interim Vice President for Health Sciences at the University of Florida.

The following papers are representative of his publications:

Linden, R. M., and K. I. Berns. 1997. Site-specific integration by adeno-associated virus: a basis for a potential gene therapy vector. *Gene Ther.* **4:**4–5.

Ralston, H. J., 3rd, D. S. Beattie, K. I. Berns, L. D. Goode, W. N. Kelley, W. J. Lennarz, L. J. Marton, A. J. Prange, Jr., and A. C. Shipp. 1996. Capturing the promise of science in medical schools. *Acad. Med.* **71:**1314–1323.

Linden, R. M., P. Ward, C. Giraud, E. Winocour, and K. I. Berns. 1996. Site-specific integration by adeno-associated virus. *Proc. Natl. Acad. Sci. USA* **93:**11288–11294.

Berns, K. I., and C. Giraud. 1996. Biology of adeno-associated virus. *Curr. Top. Microbiol. Immunol.* **218:**1–23.

Linden, R. M., E. Winocour, and K. I. Berns. 1996. The recombination signals for adeno-associated virus site-specific integration. *Proc. Natl. Acad. Sci. USA* **93:**7966–7972.

Giraud, C., E. Winocour, and K. I. Berns. 1994. Site-specific integration by adeno-associated virus is directed by a cellular DNA sequence. *Proc. Natl. Acad. Sci. USA* **91:**10039–10043.

Hong, G., P. Ward, and K. I. Berns. 1992. *In vitro* replication of adeno-associated virus DNA. *Proc. Natl. Acad. Sci. USA* **89:**4673–4677.

Berns, K. I. 1990. Parvovirus replication. *Microbiol. Rev.* **54:**316–329.

Kotin, R. M., M. Siniscalco, R. J. Samulski, X. D. Zhu, L. Hunter, C. A. Laughlin, S. McLaughlin, N. Muzyczka, M. Rocchi, and K. I. Berns. 1990. Site-specific integration by adeno-associated virus. *Proc. Natl. Acad. Sci. USA* **87:**2211–2215.

Hauswirth, W. W., and K. I. Berns. 1977. Origin and termination of adeno-associated virus DNA replication. *Virology* **78:**488–499.

ERIK D. A. DE CLERCQ

A Crusade for Drugs to Conquer Viruses

Why did I choose to attend medical school? Actually, because they had so much chemistry in the first (undergraduate) year and, later on, chemistry, and particularly biochemistry and medicinal chemistry, would remain my favored disciplines. I first obtained a medical degree but then did research on the mechanism of the antiviral activity of synthetic polyanions for my doctoral degree. My thesis adviser was Professor P. De Somer, then President of the University and Director of the Rega Institute for Medical Research, which he had founded in 1954. I started my research career at the Rega Institute as a young student in 1964 and I have remained there, except for several visiting fellowships, for the rest of my scientific career.

Originally I was appointed in 1966 as a research assistant of the Belgian National Funds for Scientific Research at the Rega Institute for Medical Research under Professor De Somer. But after two years in that position, I had the opportunity to spend two years, first as Lilly International Fellow, then as Damon Runyon Cancer Research Fellow at Stanford University School of Medicine under Professor T. C. Merigan. My stay at Stanford definitely oriented my career toward medical research.

On my return in 1970 to Leuven, I entered an academic career, first as docent (assistant professor), then as professor, and eventually as an "old-fashioned" professor, at one of the oldest and most "classical" European universities. I am charged with a number of courses for the second and third undergraduate years (i.e., biochemistry, microbiology, and virology) at the Leuven University School of Medicine and Dentistry. I teach an average of five hours per week (150 hours of teaching per year). Students' examinations amount to more than 500 per year—about 50% oral and 50% written. Teaching is a most rewarding experience—I feel I will never give up until I am forced to retire (at the age of 65, as is mandatory in Belgium).

But why did I pursue a scientific career? Why not just be a medical doctor, treating patients and making money? Honestly spoken, to find a "cure" for cancer, although, as it turned out, my major efforts went into the search for antiviral agents and, when AIDS came along, also in the search for anti-HIV agents. The research has proved extremely productive. During the last thirty years I have given well over 800 lectures at international congresses, conferences, and symposia at foreign universities and research institutes on various aspects of antiviral chemotherapy, including anti-HIV chemotherapy.

Under my leadership, the Rega Institute, which is to celebrate its fiftieth anniversary in 2004, has grown to be a world-renowned center for research on nucleoside analogues, antivirals, anti-HIV drugs, cytokines, and chemokines. Today it is one of the world's leading academic centers for antiviral research. While Professor De Somer focused on vaccine, interferon, and cytokine research, I brought in the chemistry as a new dimension in the search for new strategies to combat viral infections.

Initially my goal was to tackle herpes simplex virus (HSV) and vesicular stomatitis virus (as representatives of the DNA and RNA viruses, respectively), but this was soon extended to various other viral pathogens such as vaccinia virus (as representative of the poxvirus family), varicella-zoster virus, cytomegalovirus (CMV), HIV, polyomaviruses and papillomaviruses, hepatitis B virus, and hemorrhagic fever viruses, so that at present virtually all important human viral pathogens, including all eight human herpesviruses, are incorporated into our research program. The program is aimed at the molecular design, chemical synthesis, and biological evaluation of new antiviral substances; elucidation of their molecular target of action; and preclinical assessment of their efficacy and safety, thus identifying new leads for the chemotherapy of viral diseases.

My lifelong efforts have played a key role in the discovery and development of the compounds that should lead to the "cure" and are now widely used throughout the world for the treatment of virtually all of these important viral infections: valacyclovir and brivudin (BVDU) for the treatment of HSV and varicella-zoster virus (VZV) infections; cidofovir (HPMPC) for the treatment of CMV, Epstein-Barr virus (EBV), human papillomavirus (HPV), polyomavirus, adenovirus, and poxvirus infections; adefovir (PMEA) and tenofovir (PMPA) for the treatment of HIV and hepatitis B virus (HBV) infections; stavudine (d4T) for the treatment of HIV infections; and the non-nucleoside reverse transcriptase inhibitors, such as the TIBO and HEPT derivatives, for the treatment of HIV-1 infections. My work also has laid the foundation for the use of the virus-cell fusion inhibitors (such as the bicyclams) to prevent the progression to AIDS, of the *S*-adenosylhomocysteine hydrolase inhibitors (such as 3-deazaneplanocin A) to block hemorrhagic fever virus infections (such as Ebola) and poxvirus infections, and of the acyclic nucleoside phosphonates (HPMPC, PMEA) to treat virus-associated cancers such as cervical carcinoma (caused by HPV), hepatocellular carcinoma (caused by HBV), and nasopharyngeal carcinoma (caused by EBV).

In the course of my career, I have pursued various research interests, including chemotherapy of virus infections, chemotherapy of AIDS, chemotherapy of malignant diseases, molecular mechanism of action of antiviral and antitumor agents, enzyme targets for antiviral and antitumor agents, nucleoside and nucleotide analogues, reverse transcriptase inhibitors, virus adsorption, fusion and uncoating inhibitors, gene therapy via the virus-encoded thymidine kinase, tumor cell differentiation inducers, inhibitors of angiogenesis, and inducers of apoptosis of tumor cells and virus-infected cells.

Among the milestones in my crusades against viral infections, AIDS, and cancer were my 1970 description of several original inducers of interferon and resolution of their structural requirements for interferon induction; the 1974 description of the principles for reversible denaturation of interferon that ultimately helped elucidate the primary structure of interferon and its expression and cloning through DNA recombinant technology; the 1979 discovery of the antiretroviral reverse transcriptase activity of suramin, which prompted the explo-

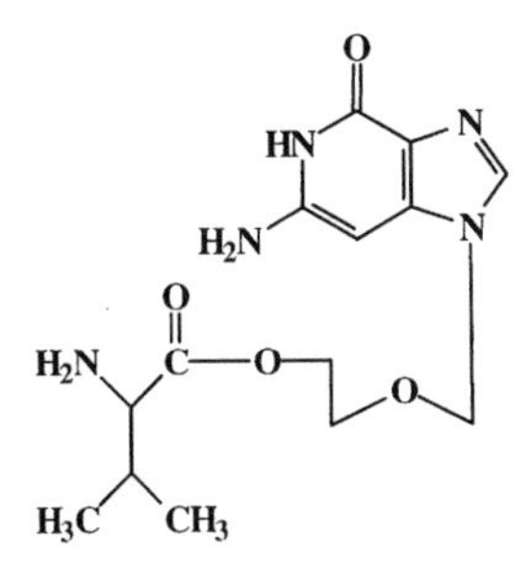

Valacyclovir
Valtrex®
Zelitrex®

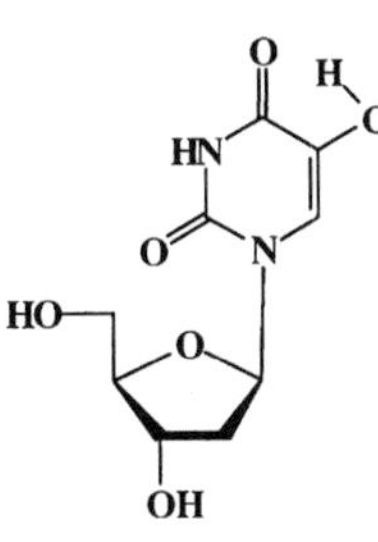

BVDU
Brivudin
Helpin®

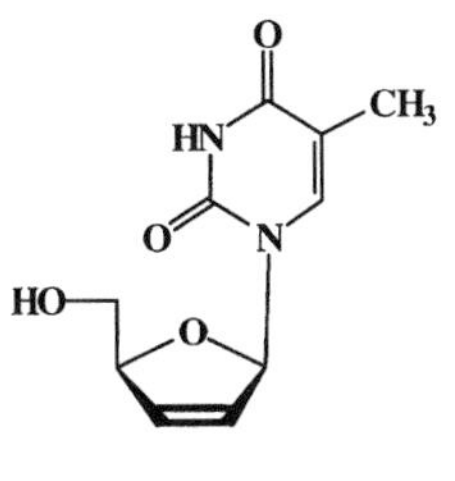

D4T
Stavudine
Zerit®

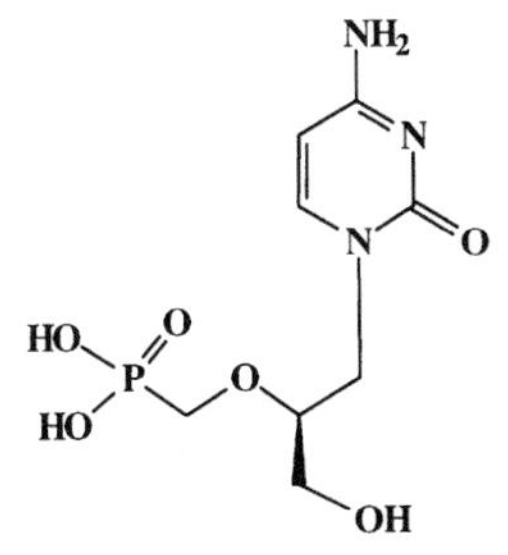

HPMPC
Cidofovir
Vistide®

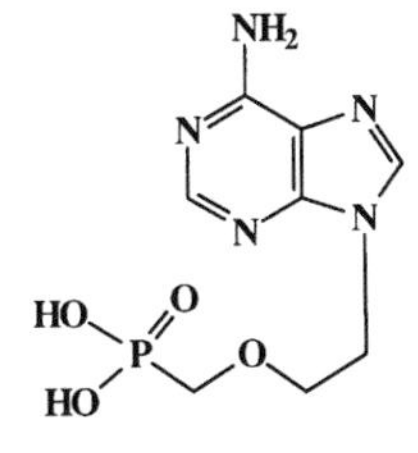

PMEA
Adefovir

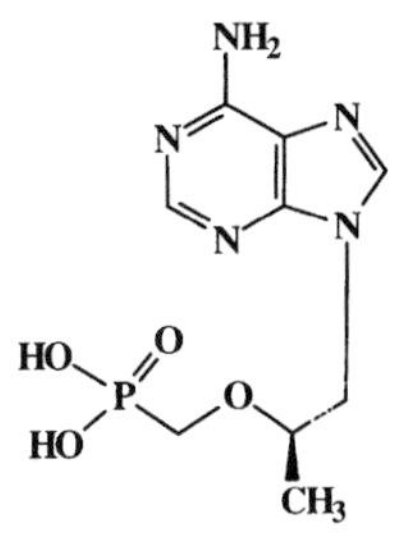

PMPA
Tenofovir

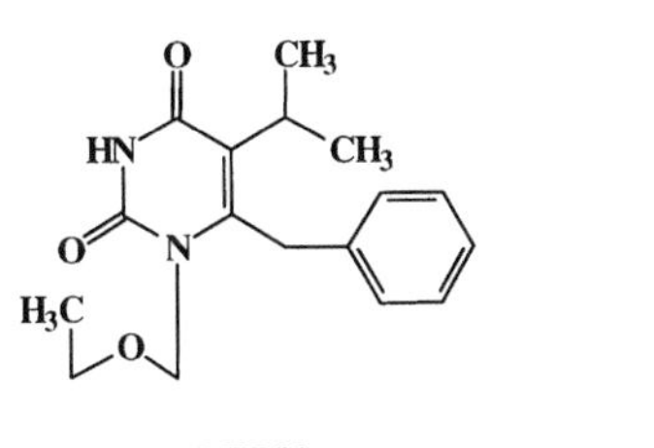

I-EBU
MKC-442
Emivirine
Coactinon™

Bicyclam
AMD3100

Selected antiviral compounds first described by Erik D. A. De Clercq.

ration of suramin as a potential anti-HIV compound; the 1978 description of the antiviral activity of dihydroxypropyladenine (DHPA), the prototype of a new class of broad-spectrum antiviral agents targeted at the S-adenosylhomocysteine hydrolase; the 1979 description of BVDU as a highly selective inhibitor of herpesviruses; the 1980 description, in collaboration with W. Fiers, J. Content, and their colleagues, of the cloning and expression of human β-interferon, a cytokine that is widely used in clinical medicine, particularly in the treatment of multiple sclerosis; the 1983 description of the synthesis and antiviral properties of aminoacyl esters of acyclovir, which led to the development of the valyl ester (valacyclovir) for the oral treatment of HSV and VZV infections; the 1985 description of the principle of combined gene therapy/chemotherapy of cancer by using antiviral agents such as ganciclovir and BVDU that are highly and specifically cytostatic to tumor cells transfected by the HSV-1 (HSV-2 or VZV) thymidine kinase gene; the 1986 discovery, in collaboration with A. Holý, of the acyclic nucleoside phosphonates, which have yielded different congeners [i.e., cidofovir (HPMPC), adefovir (PMEA), and tenofovir (PMPA)] that have proved efficacious in the treatment of herpes (i.e., HSV, VZV, CMV, EBV), HPV, polyomavirus, adenovirus, poxvirus, retrovirus (i.e., HIV) and HBV infections; the 1987 description of the anti-HIV activity of several dideoxynucleoside analogues, including d4T (stavudine), which is one of the most successful antiviral drugs used (in various drug combination schedules) in the treatment of HIV infections; the 1992 discovery of the bicyclams as an entirely new class of highly potent and selective HIV inhibitors that are specifically targeted at the virus fusion/uncoating process [these were the first compounds shown (in 1997) to block HIV infection through a specific antagonism with CXCR4, the coreceptor for T-lymphotropic X4 HIV strains; AMD3100, the prototype of the bicyclams, is the first of the receptor antagonists to enter clinical trials (in 1999) for the treatment of HIV infections]; the 1996 description of the principles for the "cure" of AIDS, based on the combination of different anti-HIV drugs that, when used from the beginning (i.e., as soon as possible after the infection) at sufficiently high ("knocking-out") concentrations, can completely suppress HIV replication and thus prevent drug-resistant virus strains from emerging; and the 1997 description of the considerable potential of acyclic nucleoside phosphonates, such as HPMPC and PMEA, as anticancer agents, through induction of tumor cell differentiation (PMEA), induction of apoptosis (HPMPC, PMEA), and inhibition of angiogenesis (HPMPC) (this may help to explain the complete and durable regression of papillomatous lesions that has been achieved by HPMPC in humans).

Such a wide array of interests requires collaboration, and I am fortunate to have at the Rega Institute a dedicated group of more than fifty coworkers, all sharing my research interests in the search for an effective cure for viral infections, AIDS, and cancer. But even such a large group is not enough to carry forth the effort of finding antiviral agents to combat emerging viral infections. Although there are numerous antimicrobials for treating bacterial infections, we still have a long way to go to combat viral pathogens, despite our successes. Hence, I have set up a collaborative worldwide network, with more than two hundred laboratories and more than one thousand individuals, mainly chemists. The collaborative centers are located in five continents and help to establish and maintain our leading position in the field. Their dedication and confidence in our work have been paramount in our crusade toward the conquest of viruses and the discovery of many antiviral compounds.

ERIK D. A. DE CLERCQ was born on March 28, 1941, in a small Flemish city, Dendermonde (in Belgium). He grew up in Hamme, approximately 10 km from Dendermonde. After high school (at the Heilige Maagd College) in Dendermonde, he studied medicine at the Katholieke Universiteit Leuven, where he graduated in 1966 summa cum laude. He received his PhD degree from Leuven University in 1972. He was appointed Professor at the Medical School of the University of Leuven in 1977 and served as Chair of the Department of Microbiology at Leuven University from 1986 to 1991. In 1985 he was elected Chair of the Directory Board of the Rega Foundation and in 1986 he was elected Chair of the Directory Board of the Rega Institute in Leuven. He still holds these offices. In 1999 he was again elected Chair of the Department of Microbiology and Immunology of the University of Leuven Medical School for a five-year term. He has served as a Wellcome Visiting Professor in Microbiological Sciences at the University of South Carolina in 1990, as a Francqui Chair Professor at the University of Antwerp, Belgium, in 1991, as Sir Henry Hallett Dale Visiting Professor at Johns Hopkins School of Medicine in 1998, and as Honorary Lecture Chair of the National Council of Taiwan, also in 1998. Since 1995 he has held the Professor P. De Somer Chair for Microbiology at the University of Leuven. He was one of the founders of the International Society for Antiviral Research and served as its President between 1990 and 1992. He received the Hoechst Marion Roussel Award of the American Society for Microbiology in 1996 and the Award for Excellence in Scientific Research of the International Society for Antiviral Research in 1998. He has been awarded honorary doctoral degrees by the Universities of Ghent and Athens and the Czech Republic Academy of Sciences. His most tangible, and most rewarding, contributions are his discoveries that have led to a panoply of new antiviral drugs that have met medical needs and helped alleviate the symptoms, if not saved the lives, of patients suffering from the most severe and often life-threatening viral infections.

The following papers are representative of his publications:

De Clercq, E., N. Yamamoto, R. Pauwels, M. Baba, D. Schols, H. Nakashima, J. Balzarini, Z. Debyser, B. A. Murrer, D. Schwartz, D. Thornton, G. Bridger, S. Fricker, G. Henson, M. Abrams, and D. Picker. 1992. Potent and selective inhibition of human immunodeficiency virus (HIV)-1 and HIV-2 replication by a class of bicyclams interacting with a viral uncoating event. *Proc. Natl. Acad. Sci. USA* **89:**5286–5290.

Pauwels, R., K. Andries, J. Desmyter, D. Schols, M. J. Kukla, H. J. Breslin, A. Raeymaeckers, J. Van Gelder, R. Woestenborghs, J. Heykants, K. Schellekens, M. A. C. Janssen, E. De Clercq, and P. A. J. Janssen. 1990. Potent and selective inhibition of HIV-1 replication *in vitro* by a novel series of TIBO derivatives. *Nature* **343:**470–474.

Baba, M., H. Tanaka, E. De Clercq, R. Pauwels, J. Balzarini, D. Schols, H. Nakashima, C.-F. Perno, R. T. Walker, and T. Miyasaka. 1989. Highly specific inhibition of human immunodeficiency virus type 1 by a novel 6-substituted acyclouridine derivative. *Biochem. Biophys. Res. Commun.* **165:**1375–1381.

Baba, M., R. Pauwels, P. Herdewijn, E. De Clercq, J. Desmyter, and M. Vandeputte. 1987. Both 2',3'-dideoxythymidine and its 2',3'-unsaturated derivative (2',3'-dideoxythymidinene) are potent and selective inhibitors of human immunodeficiency virus replication *in vitro*. *Biochem. Biophys. Res. Commun.* **142:**128–134.

De Clercq, E., A. Holy, I. Rosenberg, T. Sakuma, J. Balzarini, and P. C. Maudgal. 1986. A novel selective broad-spectrum anti-DNA virus agent. *Nature* **323:**464–467.

De Clercq, E., J. Descamps, P. De Somer, P. J. Barr, A. S. Jones, and R. T. Walker. 1979. *E*-5-(2-Bromovinyl)-2'-deoxyuridine: a potent and selective antiherpes agent. *Proc. Natl. Acad. Sci. USA* **76:**2947–2951.

Section Seven

Molecular Revelations

As the primitive microbe hunter Antony van Leeuwenhoek had done three centuries before, pioneering scientists of the twentieth century sought to see further into the mysteries of life—seeking the holy grail of heredity. And while most sought proteins, querying whether each held the secrets of life, Canadian-born, physician-trained bacteriologist Oswald Avery shocked the scientific world by showing that transformation of pneumococci was caused by a nucleic acid. By exposing the chemical that could transform harmless bacteria into deadly pathogens, the mild-mannered, charmingly courteous Avery set the stage for a most amazing show, a mystery enacted by molecules that would play out the secret of inheritance—revealing all. "The gene in solution, one is tempted to call it—appears to be nucleic acid of the desoxyribose type. Whatever it be, it is something which should be capable of complete description in terms of structural chemistry." So predicted Sir Henry Dale in his 1946 presidential address to the Royal Society. His was a call to action—a call for a new Pasteur to come forward and reinvigorate the biological sciences, to solve the myths of inheritance, to find the basis of life long hidden within the living cell, to reveal the messianic molecule. And that is exactly what James Watson and Francis Crick did. These unlikely collaborators—Watson, the restless American biologist who had traveled to Cambridge to seek scientific immortality at the age of 23, and Crick, the immodest 34-year-old British physicist, with his quick and penetrating mind—won the race. Together they gave birth to the new age of molecular biology.

In *The Double Helix*, Watson remembered later: "I huddled next to the fireplace, daydreaming about how several DNA chains could fold together in a pretty and hopefully scientific way. Soon, however, I abandoned thinking at the molecular level and turned to the much easier job of reading biochemical papers on the interrelations of DNA, RNA, and protein synthesis. . . . Virtually all the evidence then available made me believe that DNA was the template upon which RNA chains were made. In turn, RNA chains were the likely candidates for the templates for protein synthesis. . . . The idea of the genes' being immortal smelled right, and so on the wall above my desk I taped up a paper sheet saying DNA→RNA→protein. The arrows did not signify chemical transformations, but instead expressed the transfer of genetic information from the sequences of nucleotides in DNA molecules to the sequences of amino acids in proteins. . . . [w]ith the thought that I understood the relationship between nucleic acids and protein synthesis, the chill of dressing in an ice-cold bedroom brought me back to knowing truth that a slogan was no substitute for the DNA structure. . . . It came while I was drawing the fused rings of adenine on paper. Suddenly I realized the potentially profound implications of a DNA structure. . . . My pulse began to race. If this was DNA, I should create a bombshell by announcing its discovery. The existence of two intertwined chains with identical base sequences could not be a chance matter. Instead it would strongly suggest that one chain in each molecule had at some earlier stage served as the template for the synthesis of the other chain. Under this scheme, gene replication starts with the separation of its two identical chains. Then two new daughter strands are made on the two parental templates, thereby forming two DNA molecules identical to the original molecule. Thus, the essential trick of gene replication could come from the requirement that each base in the newly synthesized chain always hydrogen-bonds to an identical base."

And with the revelation of the molecular basis of life—of all life—a new generation of molecular microbe hunters set out to search further yet: to read the blueprints of each bacterium, fungus, and virus, to know the totality of a microbe's essence, to reveal the nucleic acid sequence of its entire genome, to understand the message locked within each living cell—to bring about the molecular revolution.

DAVID SCHLESSINGER

Pioneering Molecular Biology

As a very-wet-behind-the-ears sixteen-year-old high school graduate, I arrived at the College of the University of Chicago in September 1953. The natural sciences syllabus had just added the newly published *Nature* paper of Watson and Crick to its usual lineup of Newton and Mendel. The impact of the discovery of the structure of DNA on a freshman chemistry major was considerable.

By my fourth year as an undergraduate, I found that organic chemistry was the branch of chemistry that was easiest for me, and most interesting, but it was despairingly full of memorization of complex name reactions with poor yields. In contrast, the simplicity and power of the DNA model continued to be fascinating. I had also started to work in the lab of Eugene Goldwasser as a part-time technician and had done the initial steps in the purification of erythropoietin: an entire field of hormone action and chemistry opening up, among so many others! Also, I had been reading that bacteria could modify intermediary metabolites, including lactate, in many ways—and in high yields—just by using appropriate enzymes.

The attraction of biochemistry and microbiology became increasingly great, and I applied to Harvard University for graduate study. There, I was one of the first graduate students with James D. Watson.

Molecular biology is so young a set of techniques and ideas that a number of current practitioners have lived through its entire development. I was fortunate to follow much of its embryonic period in Watson's laboratory and to see "from below" the interactions of the great scientists who defined the first rash of ideas. It was easy to make discoveries and get jobs and grants in those days because everything was wide open and everyone was a raw recruit. It is a source of wistful amusement to think that I made the first pure preparations of 30S, 50S, and 70S ribosomes from *Escherichia coli* and measured their molecular weights and that my Ph.D. research also included one of the first functional in vitro systems for bacterial protein synthesis.

Results that I obtained with subcellular systems provided some of the indications that RNA was involved in directing protein formation. At the time, the notion of messenger RNA was just being formulated, in large part in the group of Jacob and Monod in Paris. Again, I was privileged to work in a postdoctoral "stage" with a remarkable group, that of Jacques Monod at the Institut Pasteur. There I realized more fully the wide-open domains that were added to microbiology by the French school, including much of microbial genetics, growth control, and scrupulous attention to the balance of physiological processes.

When I arrived at Washington University in St. Louis in 1962 as an instructor, I began an independent career by attacking a new problem: the analysis of bacterial membranes. At that time, membranes were a nonexistent field of study, and I soon found out why. After a year, the only substantive progress I made was to determine that my membrane preparations were always highly contaminated with RNA. The contaminating RNA, however, released with nonionic detergents soon proved to be ribosomes (in fact, the first bacterial polyribosomes to be observed), and I realized that RNA must be my research fate.

In other work at the time, I participated in the comparable discovery of mammalian polysomes and stable messenger RNA in reticulocytes and initiated two long-term projects in *E. coli*: (1) studies of messenger RNA turnover in subcellular systems that identified several of the enzymes involved and provided some of the early hypotheses about the control of turnover and (2) formulation and analysis of the ribosome cycle in protein synthesis. With my colleagues David Apirion and Giorgio Mangiarotti, I analyzed the dynamics of ribosome metabolism, facilitating the study with fragile mutants of *E. coli* that could be lysed gently enough to preserve the polysomal structures.

The formulation of the ribosome cycle was extended to the analysis of the action of antiribosome antibiotics, and with Lucio Luzzatto, I discovered the specific block of polysome function at initiation that explains the bactericidal action of streptomycin. That work led to the recognition of the Eli Lilly Award in Microbiology presented by the American Society for Microbiology.

Throughout the next decades, as I continued to teach and conduct research in microbiology, gradually rising in the professorial ranks, I sustained an interest in infectious diseases, which centered on antibacterial and antifungal agents. Parallel work on nucleic acid metabolism followed my discovery with Nikolai Nikolaev of the role of double-stranded ribonucleases in the formation of mature RNA species and subsequent analyses of ribosome formation.

In recent years, I have continued the adroit choice of collaborators who were initiating pioneering ventures. In 1987, an increasing interest in long-range chromatin organization led me to join forces with Maynard Olson and my long-term associate Michele D'Urso in the development and exploitation of yeast artificial chromosome technology for the Human Genome Project. Although X chromosome mapping and technology development for gene searches may seem a far cry from traditional microbiology, it can be recalled that cell biology, genome mapping, and biotechnology are all disciplines that derive from classical microbiology, and they continue to depend on classical microbiology. One can note that the human genetics community would have found it unlikely that genome studies would become essentially totally dependent on the use of yeast hosts, clones, and genetics—in addition to the more traditional bacterial systems that already dominated the approaches to positional cloning.

After almost forty-five years of research, working with new ideas and students who become colleagues remains fun. Genome analysis and the study of X-linked diseases provide the current focus of my work, but I have maintained an avid interest in all branches of microbiology. The great renaissance of "real" microbiology is now just beginning, with the application of genome approaches to topics of the greatest scientific and practical interest in the understanding of evolution by the comparative analysis of microbial biochemistry, the use of microbial agents to solve environmental pollution problems, and the analysis of microbial pathology to conquer infectious diseases.

The resurgence of microbiology and the parallel extension of powerful genomic approaches are in full swing. The expansion of microbial genomics in the nascent century will include not only (1) the exploration of the fraction of genome space that has been selected for genes during evolution but also (2) the use of combinatorial peptide and RNA chemistry to investigate "genes" and "motifs," like alternate ATP-binding sites, that were not chosen by evolution and (3) the study of auxiliary monomers and polymers that can increase the repertoire of enzyme catalysts and drugs beyond the metabolites that have been favored on earth.

As for my own work, involvement in discussions for the planning of microbial sequencing initiatives has been accompanied by a more interventive participation in the expansion of human genome approaches. At the Human Genome Center we had organized in St. Louis, we produced a map of the X chromosome and identified the causative genes for two X-linked diseases (anhidrotic ectodermal dysplasia and the Simpson-Golabi-Behmel gigantism/overgrowth syndrome). With the map completed we have turned to studies of the genetics of aging, establishing a Laboratory of Genetics at the National Institute on Aging in Baltimore. This is the first position I have had that permits me to build institutional infrastructure, directly fostering the careers of independent investigators. Our focus is on aging as an integrated extension of human development. The independent groups are working on projects to study genes in chromatin form (directed by Ramaiah Nagaraja and Weidong Wang) and analysis of the relation of embryonic development to the mortalization of cells (the group of Minoru Ko). We are also initiating, with Giuseppe Pilia, studies of the genetic basis of age-related phenomena in the relatively homogeneous population of Sardinia.

Thus, as one matures, the dictum of Pasteur remains firm, that research constantly opens new vistas.

DAVID SCHLESSINGER was born in Toronto, Canada, in 1936. He was educated at the University of Chicago and Harvard University. He worked at the Pasteur Institute in Paris in the early 1960s before coming to the Department of Microbiology at the School of Medicine at Washington University in St. Louis. He was President of the American Society for Microbiology from 1994 to 1995. Dr. Schlessinger studies cell physiology and biochemistry. His pioneering studies have helped elucidate the role of RNA in cellular functions. Since 1997, he has focused on the study of the genetics of aging.

The following papers are representative of his publications:

Pilia, G., R. M. Hughes-Benzie, A. MacKenzie, P. Baybayan, E. Y. Chen, R. Huber, G. Neri, A. Cao, A. Forabosco, and D. Schlessinger. 1996. Mutations in GPC3, a glypican gene, cause the Simpson-Golabi-Behmel overgrowth syndrome. *Nat. Genet.* **12:**241–247.

Sirdeshmukh, R., and D. Schlessinger. 1985. Why is processing of 23S ribosomal RNA in *Escherichia coli* not obligate for its function? *J. Mol. Biol.* **186:**669–672.

Nikolaev, N., L. Silengo, and D. Schlessinger. 1973. A role of RNase III in processing of rRNA and mRNA precursors in *Escherichia coli*. *J. Biol. Chem.* **248:**7967–7969.

Luzzatto, L., D. Apirion, and D. Schlessinger. 1968. Mechanism of action of streptomycin in *Escherichia coli*: interruption of the ribosome cycle at the initiation of protein synthesis. *Proc. Natl. Acad. Sci. USA* **60:**873–880.

Mangiarotti, G., and D. Schlessinger. 1967. Polyribosome metabolism. Formation and lifetime of messenger RNA molecules, ribosomal subunit couples, and polyribosomes. *J. Mol. Biol.* **29:**395–418.

23 SUSAN GOTTESMAN

Solving Genetic Puzzles

For me, a career in bacterial genetics has meant solving genetic logic puzzles with colored colonies or bacteriophage plaques on Petri dishes. I've been lucky enough to do that in an environment of excellent support and stimulating colleagues at the National Institutes of Health (NIH). I spent elementary school on Long Island, working my way through whatever books happened to come my way: Nancy Drew, various biographies, and, sometime in sixth grade, *Microbe Hunters*. The portrayal of the drama and science behind the first discoveries of bacteria and their role in disease clearly appealed to me, enough for an instant decision to become a microbiologist, although I'm not sure what I imagined that would mean. I found the logic of Koch's approach for proving that a particular organism caused a particular disease particularly appealing, as well as the description of figuring out how to make single colonies on a potato slice. It was the scientific ideas rather than the medical relevance that attracted me. I had the wrong personality to deal with patients, not a great memory for facts unless I could fit them into a logical theory, no stomach for the messier aspects of medicine, and little family pressure to become a physician.

Luckily, by the time I was in junior high and high school, the consequences of Sputnik were firmly in place—lots of emphasis on science and a variety of summer programs encouraging students to stay in science. The one I attended included lectures about the emerging wonder substance, DNA, a molecule that had not yet made it to my regular classroom or textbooks. The accompanying "research project" started the process of convincing me that not only did the logical construct of science entertain me, but I definitely enjoyed doing experiments as well. My group was trying to grow some bacteria on defined media, although I can't tell you what the bacteria were, and I don't think we succeeded.

I went off to Radcliffe College as a biochemical sciences major and managed to enroll in a freshman seminar in microbial genetics, this time studying phase variation in *Serratia marcescens*. By the end of freshman year, I missed labs and Petri dishes when they weren't available, sufficiently to resist the lure of courses in the social sciences and to take a part-time job for a couple of years as one of a crowd of "Cliffie" technicians in J. D. Watson's lab. I remember being interviewed for the position by Watson. It was brief and probably was the only direct interaction I had with him. I recall him telling me that it was unlikely I would get to do any experiments of my own, at least not until I was in graduate school. That didn't bother me; I didn't have a burning question I wanted to answer and was perfectly happy just to be around a lab. It really didn't matter whose lab as long as I could get my hands wet in science. Two other Cliffies and I were all helping Gary Gussin finish his thesis work, doing innumerable trichloroacetic acid precipitations onto filters on planchets for radioactive counting. Every afternoon at four o'clock everything would cease for afternoon tea, a habit Watson had brought back from his years in England. This was a fairly stratified affair, with the technicians and secretaries at a table on the side, and the students and postdocs at the major table. I suppose if I had been a bit brasher, I could have sat at the table with Watson, but I'm not sure it occurred to me. Even so, the students that were in the lab then have remained friends and colleagues today.

While still an undergraduate, I met my husband Michael in a physics lab; we were both biochemistry majors. We married in my junior and his senior year in college. Michael went on to become a physician and biomedical researcher. He had all the right traits for a medical career that I was, and still am, missing. My own orientation to science and to microbes was reinforced by a senior thesis on the kinetics of induction of the *lac* operon in the MIT laboratory of my biochemical sciences tutor, Boris Magasanik. My career path was set.

While we stayed in the Boston area for graduate school (Michael as a medical student), I moved across the river from Cambridge and went to the Microbiology and Molecular Genetics Department at Harvard Medical School to work with Jonathan Beckwith. The project was to isolate, before the advent of cloning, a specialized transducing phage carrying the *ara* operon to allow the identification and characterization of the AraC gene product, suspected to be a positive regulator. I didn't get to the last objective, but the four years there introduced me to the use of two kinds of toothpicks to pick colonies and the genetic approaches pioneered in the Beckwith lab that I have used since.

In 1971 we headed for NIH for postdoctoral study, with our one-and-a-half-year-old son in tow. For Michael, a three-year research associate appointment in the Public Health Service was an alternative to the Vietnam war draft; for me it was a chance to work on bacteriophage lambda in the midst of an intense and exciting bacteriophage group, adding the phage expertise to my bacterial genetics training. I worked with Max Gottesman who was (and still is) unrelated to Michael, although the degree of confusion wasn't helped by our all originally occupying the same building, similar initials, and a shared interest in *Escherichia coli* (Michael was then working with Marty Gellert on *E. coli* ligase). Eventually there was a paper referred to as "Gottesman cubed": Gottesman (Michael), Gottesman (Max), Gottesman (Susan), and Marty Gellert. I did enough mapping of the locus to earn my authorship, and the hope was that this paper would prove that there were three separate people, and not one very busy one. Confusion still abounds but is generally more entertaining than bothersome.

My postdoctoral project was on the site-specific excision of bacteriophage lambda. Mark Shulman, who had preceded me in the lab, had developed a lambda derivative that allowed an intramolecular excision reaction. The aim was to reconstitute the reaction in vitro; Howard Nash was using similar approaches to look at the integration reaction. Some progress was made, and we found out something about the in vivo reaction as well. Along the way our daughter was born.

After three years at NIH, we returned to Boston so that Michael could finish a residency in internal medicine. David Botstein, who I had met when he was on my qualifying exam committee, offered me support as a research associate in his lab at MIT, with the freedom to continue my own work while we decided where we would end up. It was a good place to be and a good time to be there. Nancy Kleckner was a postdoc in the lab studying the Tn10 transposon and the tricks it could do. With a high school student, we isolated one of the first sets of Tn10 insertional auxotrophs that we still use frequently for mapping and strain construction.

I was continuing to work on reconstructing the lambda excision reaction in vitro. Xis, a protein necessary for the reaction, was known to be functionally unstable in vivo. When David Zipser's lab at Cold Spring Harbor identified some "*deg*" mutants (defective in degradation of fragments of β-galactosidase) and showed that one of them was in the pleiotropic *lon* mutant, I requested the mutant strains to find out if they would help in producing Xis by stabilizing it. The *lon* mutants were also sensitive to ultraviolet light and overproduced capsular polysaccharide; the basis for these phenotypes was not known. To test the *deg* mutations, I needed to move them into fresh strains and wanted a simple plate assay to follow the defect in protein degradation. It seemed possible that degradation, if it destroyed protein fragments, might also destroy some proteins carrying missense mutations. If so, a mutant in the degradation system might suppress the missense mutation. The simplest way to assay for suppression of a missense mutation would be if the mutation was carried on a phage; the phage might fail to plaque unless the *deg* system was mutant. I wandered the halls of MIT collecting phage missense mutations, mostly temperature-sensitive (*ts*) mutations in essential functions. Nancy Kleckner had a set of lambda *ts* mutants; Jon King had various T4 late gene *ts* mutants. The collection was tested on my deg^+ and deg^- candidates at various temperatures. As I had hoped, a number of the *ts* mutants plated on the deg^- strain at temperatures that were nonpermissive on the deg^+ strain. This provided a simple and trustworthy assay for tracking the Deg phenotype of *lon* mutants and told me that some of the *ts* mutants were likely to be *ts* only because they were degraded at the elevated temperatures. The *lon* mutants didn't lead to much Xis stabilization, but by then I was interested enough in Lon's role in the cell to begin to develop protein degradation as my major research interest.

While I was at MIT, the first guidelines for recombinant DNA were being developed, and Nancy Kleckner and I sat on a committee developing "biological containment" guidelines. My tendency to voice my strong opinions developed into continued participation in the Recombinant DNA Advisory Committee (RAC) for the next ten years—a chance to learn more than I sometimes wanted about organisms other than *E. coli*, to meet a variety of other scientists, and to learn a bit about the intersection of science with some of the wilder aspects of science policy. Having worked in the politically active Beckwith lab, I had had an opportunity to think about some of the broader implications of recombinant DNA. Day to day, however, developing and interpreting the guidelines was often like an abstruse game in which the aim was to interpret the written rules to fit the desired outcome. Keeping track of the underlying scientific logic was a challenge. Sitting on RAC also served as an instant introduction to the problems of translating scientific issues for the general public, usually via the press. At one point, we were considering public comments on one of the major revisions; we had gotten large numbers of letters, some of them a bit strange. I read an excerpt from one, worrying about whether scientists were going to be creating pigs with wings; later it appeared in *Time* magazine in a compilation of strange quotes of the year—with my name appended.

Both Michael and I were offered jobs back at NIH in 1976 by Ira Pastan, in the Laboratory of Molecular Biology at the National Cancer Institute, where I had been a postdoc. Having two real jobs in one place that we enjoyed was hard to match. We've never found a good reason to leave. Although we've only had two papers together over the years (one in 1974, the next in 1993), the common research lifestyle more than the common subject interest was important. We might have argued over who should stay home with a sick child and who gets to go in and do an experiment, or complained about predicted departure times that stretch on for extra hours, but we each understood the other's degree of involvement and excitement when things were happening at work.

While I continued working on site-specific recombination for a number of years, my other project, untangling the complicated genetics of the Lon system, occupied an increasing amount of my attention. We wanted to understand what sort of cellular substrates Lon degraded and how proteolysis is used by the cell to regulate gene expression. Suppressors that only reversed a single of the multiple *lon* phenotypes supported the idea that the multiple phenotypes were due to multiple unstable substrates and led to identification of the likely *lon* targets. The ultimate test was the demonstration that our suspected targets were rapidly degraded but were stabilized in *lon* mutants.

In 1983 Michael Maurizi, who had trained with Bob Switzer at the University of Illinois and Earl Stadtman at NIH, joined the lab and began a long-term collaboration that added in vitro analysis of the protease and its action to our interests. Most important, he began a search for proteases other than *lon* that we knew lurked in *E. coli*. His discovery and subsequent analysis of the ClpAP protease and the related ClpXP protease provided the first evidence of a well-conserved new family of energy-dependent proteases with critical substrates in many bacterial systems. Although our work continues to focus on *E. coli*, a variety of information has indicated that the Clp proteases are strikingly similar in architecture and function to the eukaryotic 26S protease, responsible for degradation of cyclins and a growing number of other regulators. I continue to be most entertained if we can analyze a complicated regulatory circuit by the color of colonies on indicator plates; my collaborators at NIH, Michael Maurizi and Sue Wickner, consider it equally entertaining to look at the purified proteases in vitro. The result has been a constantly stimulating and active research environment, in the utopia of funding without grants that NIH provides.

SUSAN GOTTESMAN was born in New York and raised on Long Island. As an undergraduate at Radcliffe College, she did research on induction kinetics of β-galactosidase under the tutorship of Boris Magasanik. Her doctoral degree was awarded in 1972 in the Microbiology and Molecular Genetics Department at Harvard, where she worked under the mentorship of Jonathan Beckwith on the isolation of a specialized transducing phage by a generalizable method. She was a postdoctoral fellow, supported by the Jane Coffin Childs Memorial Fund for Cancer Research, at the National Institutes of Health (NIH) with Max Gottesman, where she worked on site-specific recombination in lambda. She went on to become a research associate at MIT in the laboratory of David Botstein, continuing an independent project on site-specific recombination; there she also began genetic studies on proteolysis with work on the role of *lon* mutants in suppressing some missense mutants. She returned to NIH in an independent research position, continued studies on site-specific recombination, and expanded studies to look at the basis for the multiple phenotypes of *lon* mutants. She served on the Recombinant DNA Advisory Committee (RAC) for a ten-year tenure. She then went on to identify a novel two-component ATP-dependent protease, ClpAP, which proved to be the first of a highly conserved family of proteins, possessing both chaperone activity and protease activity. Her work on proteolysis was recognized with election to the National Academy of Sciences in 1998.

The following papers are representative of her publications:

Gottesman, S., S. Wickner, and M. R. Maurizi. 1997. Protein quality control: triage by chaperones and proteases. *Genes Dev.* **11:**815–823.

Sledjeski, D. D., A. Gupta, and S. Gottesman. 1996. The small RNA, DsrA, is essential for the low temperature expression of RpoS during exponential growth in *Escherichia coli*. *EMBO J.* **15:**3993–4000.

Wickner, S., S. Gottesman, D. Skowyra, J. Hoskins, K. McKenney, and M. Maurizi. 1994. A molecular chaperone, ClpA, functions like DnaK and DnaJ. *Proc. Natl. Acad. Sci. USA* **91:**12218–12222.

Maurizi, M. R., W. P. Clark, S.-H Kim, and S. Gottesman. 1990. ClpP represents a unique family of serine proteases. *J. Biol. Chem.* **265:**12546–12552.

Gottesman, S., C. Squires, E. Pichersky, M. Carrington, M. Hobbs, J. S. Mattick, B. Dalrymple, H. Kuramitsu, T. Shiroza, T. Foster, W. P. Clark, B. Ross, C. Squires, and M. R. Maurizi. 1990. Conservation of the regulatory subunit for the Clp ATP-dependent protease in prokaryotes and eukaryotes. *Proc. Natl. Acad. Sci. USA* **87:**3513–3517.

Torres-Cabassa, A. S., and S. Gottesman. 1987. Capsule synthesis in *Escherichia coli* K-12 is regulated by proteolysis. *J. Bacteriol.* **169:**981–989.

Mizusawa, S., and S. Gottesman. 1983. Protein degradation in *Escherichia coli*: The *lon* gene controls the stability of SulA protein. *Proc. Natl. Acad. Sci. USA* **80:**358–362.

An Accidental Microbiologist

I don't ever remember not having an absolute love of learning new things. This was probably due, in part, to both of my parents' being educators (my father was a high school principal and my mother was an elementary school teacher), and it was always expected that my brother and I would approach our education with the utmost seriousness and dedication. I grew up in a middle-class family in the 1960s and early 1970s in Saugus, Massachusetts, a northern suburb of Boston, where I attended public school. It wasn't until I was a freshman in high school that I became fascinated with science. I was extremely fortunate to have been taught biology that year by a young, energetic teacher, John Collins, who had the ability to make science come alive. This was followed by an extraordinary year of chemistry taught by Harold Everett, a retired chemist from DuPont. At this point in my life I was convinced that I wanted a career grounded in the biological sciences, but the only way that I knew how to achieve that dream was to go on to practice medicine. I had never met anybody with a career in research, and so without such a role model I set off on a path for medical school.

When it came time to select an institution for undergraduate study I looked at several colleges and universities in the Northeast. Throughout high school I had felt somewhat out of place being female and being interested in science. I enrolled in Rensselaer Polytechnic Institute because I felt so at ease there, given that essentially all of the students were majoring in science or engineering. Of course, the male to female ratio of 11 to 1 didn't hurt either. In addition to the required courses in biology, organic and physical chemistry, and physics, I took other science courses, including physiology, immunology, virology, and microbiology. As a senior, I took a special research projects course to work in the laboratory of Lenore Clesceri with her postdoctoral fellow, George Pierce. We isolated bacteria from Lake George (located just north of the campus) that had the ability to degrade herbicides in the hope that we would find a means of combating the pollution that was contaminating the lake in runoff from the surrounding farmland. Of all the science that I had been exposed to by this point, this was by far the most exciting of my life. I spent almost every free moment in the laboratory that year. I soon recognized that I was far more interested in a research career than a medical one. This was disconcerting because I had been applying to and interviewing with medical schools during this time, having set my sights on this career path for eight years. At the very last minute, I began a search for a graduate program. The geographical search was narrowed considerably by the fact that my boyfriend was working in Toronto, and so in the fall of 1977 I began graduate school in pharmacology at the State University of New York at Buffalo, which was the closest point in the United States to Toronto.

Once I began graduate studies I knew that I had made the right choice. After several research rotations in laboratories in the department, I began thesis work in the laboratory of a young assistant professor, Dr. J. Craig Venter. His laboratory was working on β-adrenergic receptors, and although we know today that they are part of a much larger family of receptors that mediate signal transduction via G proteins, there was very little molecular information on these receptor proteins in the late 1970s. I was very excited about the potential of using monoclonal antibodies, which had been described only two years earlier, to probe structural differences between β_1- and β_2-adrenergic receptors. One of the most important lessons that I learned from Craig as a mentor is not to be afraid of new approaches. Therefore, without any formal training in tissue culture or immunology, and despite the pessimism of a member of my thesis committee who thought this approach would never work, I set out to create receptor-specific antibodies. Several months after finishing my thesis work, Craig and I were married and began our long-term scientific collaboration.

My interest in the receptor field continued to grow. Craig and I moved to the National Institute of Neurological Disorders and Stroke (NINDS) at the National Institutes of Health (NIH) in 1984, where we had the opportunity to learn molecular biology. A race was on among a few labs around the world to see who would be the first group to successfully clone a G protein–coupled receptor. We were in the final stages of cloning a β-adrenergic receptor from human brain when Dr. Robert Lefkowitz' lab at Duke reported the sequence of a hamster β-adrenergic receptor. To everyone's surprise, the secondary structure of the β-adrenergic receptor was similar to that of rhodopsin, the retinal protein that serves as a receptor for light. We finished our cloning work, and with renewed enthusiasm and a cloned receptor in hand, I set about developing a system for its stable expression in cultured cells. These were exciting times in the receptor field as new G protein–coupled receptors were cloned and as new insights into the molecular mechanisms of these proteins continued to accumulate.

My only disappointment during this period was that it had become clear that after five years in NINDS, I was unlikely to get tenure unless I could show my ability as an independent researcher. In 1989, Dr. Boris Tabakoff, the Scientific Director of the National Institute on Alcohol Abuse and Alcoholism (NIAAA), offered me the opportunity to move my research group into NIAAA to head up a receptor laboratory. It was becoming clear that receptor systems in the nervous system were playing fundamental roles in the processes of addiction and tolerance to alcohol. I saw this as a wonderful opportunity to demonstrate my competence as an independent scientist, even though it meant that Craig and I would have to go through a "laboratory divorce." Our research interests had moved apart, and it was the right time for a trial separation of laboratory lives. Craig had begun work on large-scale automated DNA sequencing of human chromosomal regions, and I had focused my efforts solely on the molecular biology of receptors. Frustrated by the slow pace of manual DNA sequencing, he had established a Cooperative Research and Development Agreement with Applied Biosystems to become the first test site for their new automated DNA sequencer.

And then the opportunity to create an entirely new research institute was presented to Craig in 1992. He could move from NIH and work on large-scale sequencing of cDNA clones to discover new genes. For his sequencing Craig used cDNA clones, which meant that sequences he derived represented genes that are expressed; these sequences were termed *expressed sequence tags*, or ESTs. His technique generated a great deal of interest by venture capital groups because of its potential to uncover new gene-based therapeutic proteins like erythropoietin. Craig made it clear that he would only leave NIH if he were provided with the resources to start his own research institute. Many investors thought that this was outrageous, but Wallace Steinberg, President of Health Care Investment Corporation (HCIC), agreed to provide Craig with $70 million over a ten-year period to form a new research institute that would use the EST approach to carry out large-scale human gene discovery. The intellectual property generated in the course of this work would belong to a new biotechnology company to be established at the same time, and this company would work to turn basic biological findings into new therapeutic proteins. Thus, The Institute for Genomic Research (TIGR), a nonprofit research organization, and Human Genome Sciences, Inc. (HGS), a new genomics-based company, were created in July 1992. The magnitude and nature of this investment in science or biotechnology were unprecedented and became known in the investment community as "Wally's Folly."

Looking back at these events, I realize what a profound change the creation of TIGR was in our careers. However, at the time it seemed like just another opportunity to move research forward at a scale not possible at NIH. Although my own research program at NIAAA was flourishing, it was difficult to contemplate the new scale of biology that could potentially come from the application of the EST method to human gene discovery and not become excited about the possibilities. At this time, it was estimated that there were at least one thousand G protein–coupled receptors in humans, yet fewer than fifty had been cloned and sequenced. I saw the opportunity that TIGR presented as being one that could greatly accelerate my own research program.

And so, Craig's laboratory and my laboratory joined forces once again to build TIGR from the ground up. Most scientists spend their entire careers in an existing institution that has laboratory space and the necessary infrastructure to carry out research. However, twelve of us moved from NIH into a completely empty building and had to start from scratch to create a new research organization. This was both a wonderful opportunity and a tremendous challenge. It quickly made me realize how much I had taken for granted in all previous situations. We became adept at reading architectural blueprints, dealing with state and federal agencies to obtain licenses for use of radioactive materials and controlled substances, and understanding the myriad of regulations that govern the disposal of radioactive, chemical, and biohazardous wastes. Considering that we now have a staff of over two hundred, a new eight-acre campus in Rockville, Maryland, and an annual research budget of more than $35 million, I often look back to where we were when TIGR first took shape and appreciate how much we have accomplished in such a short period of time.

During the first two years, all the scientists at TIGR were involved to a large extent with human EST work. The EST method identified more than thirty new G protein–coupled receptors; however, these sequences were also of great interest to HGS and its partner, SmithKline Beecham (SB). This is because more than 60% of all prescription drugs on the market today are targeted to G protein–coupled receptors, and both HGS and SB saw these receptors as new targets for drug development. Because of the arrangement between TIGR and HGS, my laboratory was limited in what we could publish on the new receptors, a frustration that I had not anticipated but one that was directly responsible for my change in research focus back to microbiology.

Craig was at a meeting in Spain with Dr. Hamilton O. Smith from Johns Hopkins in late 1993 when they began talking about the work at TIGR. Ham Smith had spent his career working on *Haemophilus influenzae*, a human pathogen that causes otitis media and meningitis in children. It is an organism of historical significance in science as well, being the source from which Ham Smith first isolated restriction endonucleases, which led to his Nobel Prize in Medicine in 1978. He proposed that TIGR undertake the sequencing of the entire chromosome of *H. influenzae*. He estimated that 25,000 to 30,000 sequences of approximately 500 bp in length would be needed to complete the project. Given that TIGR was generating more than 200,000 sequences per year at that time, we were anxious to move forward to see if this was feasible. No complete genome sequence for a free-living organism had ever been deciphered, so we all realized that this would be a landmark achievement if it could be accomplished.

We began the project in April 1994 without a source of extramural funds. We were convinced that NIH would provide a grant, but as it turned out we were never able to convince the study section that the project was feasible—not even when we were well into the final closure phase. Despite the lack of NIH confidence that we would succeed, almost one year to the day that we began, the final gap in the *H. influenzae* sequence was closed and we opened a bottle of champagne to celebrate. At the American Society for Microbiology meeting held in Washington, D.C., in May 1995, Craig and Ham Smith received a standing ovation, and the work continues to be heralded as a hallmark of genomics by the microbiological community.

It was impossible to see the excitement generated by the *H. influenzae* project and at the same time to realize that my receptor program was being limited by the constraints of the TIGR-HGS relationship and not be extremely frustrated. Sequencing whole microbial genomes could be accomplished within months, yielding great scientific rewards. Thus, when the decision was made to use the shotgun strategy to sequence the genome of *Mycoplasma genitalium*, the organism with the smallest known genome of any free-living organism, I jumped at the chance to head up this project. The *M. genitalium* project was completed in less than six months and was an exhilarating experience.

With the second completed genome sequence in hand, it had become even more obvious to all of us how much biology we still did not understand, even for the simplest free-living organism. As we had observed in *H. influenzae*, more than 30% of the genes in *M. genitalium* were of unknown biological function. In 1996, TIGR completed a third genome sequence from the methanogenic archaeon *Methanococcus jannaschii*, which had been isolated from a hydrothermal vent almost two miles deep in the Pacific Ocean and which grows at an optimum temperature of 85°C. The *M. jannaschii* sequence represented the first complete sequence from any archaeal species. Together with the sequence of *Saccharomyces cerevisiae* that had recently been completed by a consortium of laboratories in Europe and the United States, we now were provided an opportunity for the first time to look at members of all three domains of life and ask how these species were related. The molecular diversity we observed among these species was remarkable, and the power of comparative genomics was becoming clear.

I do not think it is an exaggeration to say that the application of genomics technology has reinvigorated the field of microbiology. Since TIGR reported the first complete microbial genome sequence in 1995, the sequences of more than twenty bacterial and archaeal species have been published, and at least sixty other genome projects are in progress in laboratories around the world. I am proud to say that TIGR has continued to play a leading role in the field of microbial genomics, having contributed eight of the published genome sequences, with more than twenty-five other projects under way.

What have we learned from whole genome analysis of prokaryotic species? One of the most important lessons is that the diversity of metabolism, biochemistry, and genetics in the microbial world is remarkable. Even as the list of completed sequences continues to grow, the number of novel genes that are being discovered has not yet begun to diminish. A significant portion of the genome of every species studied to date represents genes that are unique to each organism. Genome sequencing suggests that lateral gene transfer may play a much bigger role than was previously appreciated in generating species diversity. The idea of a "model organism" in the microbial world clearly is not valid.

And although each completed project is a major accomplishment in itself and represents the contribution of a large number of scientists with training in microbiology, molecular biology, and bioinformatics, I prefer to view each genome sequence as a new starting point in our understanding of microbial biology. One of the great challenges and opportunities for microbial genomics in the next decades will be to use this new information to carry out biological studies on a genome-wide scale and begin to address new questions, including the control of gene expression and the interaction between proteins within microbial cells. Genomic information has the potential to provide new targets for development of vaccines and antimicrobial agents for important human pathogens and new biochemical pathways that may be exploited in industrial processes and environmental bioremediation.

Creating a biological picture of an organism from the list of its genes has represented some of the most intellectually challenging work that I have ever done in science. And with so many genes of unknown function identified from genome analysis, it is essential that scientists begin to apply a variety of methods to understand the role that these genes play in organismal biology. Some of these studies may be hypothesis driven and others not, yet they will all ultimately be important in our elucidation of microbial physiology.

It has now been almost five years since I began my adventures as an accidental microbiologist. These last few years have been the most exciting of my entire career. I hope that funding for microbial genome sequencing will continue for some time to come because we have only scratched the surface in terms of sampling the vast numbers of microbial species on this planet. Most species cannot be cultured in the laboratory, yet these species likely play some of the most important roles in the ecology of our planet. Breakthroughs in genomics technology and bioinformatics will continue to allow us to push back the frontiers of whole genome analysis. And yet, at the same time, new technologies for functional genomics present exciting opportunities to begin to study the dynamic nature of the microbial cell.

I have never once regretted my decision to pursue a career in research. I am extremely fortunate for all of the opportunities that have been presented to me, for all of the interesting colleagues who I have met, and, in a most serendipitous way, my return to the field of microbiology that first captivated me as an undergraduate student. The most important lessons that I have learned since 1980 are to always believe in your abilities to succeed and always embrace new ideas and opportunities because you never know where they may lead. As Yogi Berra said, "When you come to a fork in the road, take it." I truly believe that because of the power of genomics, the twenty-first century will be the century of biology. I can't think of a more exciting time to be a scientist!

CLAIRE M. FRASER was born in Boston in 1955. She is a graduate of Rensselaer Polytechnic Institute and received a bachelor of science degree in biology, summa cum laude, in 1977. She received a doctoral degree in pharmacology from the State University of New York at Buffalo in 1981. From 1981 to 1982 she was a postdoctoral research associate in the Department of Pharmacology and Therapeutics of the State University of New York at Buffalo. She spent eight years at the National Institutes of Health and was appointed Chief of the Section of Molecular Neurobiology at the National Institute on Alcohol Abuse and Alcoholism. She joined The Institute for Genomic Research (TIGR) in 1992 as Vice President for Research and was initially involved in studies to find differences in gene expression in human tumors and matched normal tissues and in using a genomic-based approach to understand the molecular basis for tumor development. She became the President of TIGR in 1998. She has been involved in whole genome sequence analysis of microbial genomes, leading the teams that sequenced the genomes of *Mycoplasma genitalium*, the smallest genome of any known free-living organism, and the two spirochetes, *Tre-*

ponema pallidum, the cause of syphilis in humans, and *Borrelia burgdorferi*, the organism responsible for tick-borne Lyme disease. She has over 130 publications in leading scientific journals. She has won several awards and honors in recognition of her accomplishments, including the American Cancer Society Junior Faculty Research Award in 1984, the Computerworld Smithsonian Award for innovation in information technology in 1998, the Burroughs-Wellcome Fund Visiting Scientist Professorship Award in 1999, and the Institute for Mathematics and Advanced Supercomputing Award in 1999. She also was named one of Maryland's top 100 women in 1997.

The following papers are representative of her publications (also see the TIGR Web site at http://www.tigr.org):

Nelson, K. E., R. A. Clayton, S. R. Gill, M. L. Gwinn, R. J. Dodson, D. H. Haft, E. K. Hickey, J. D. Peterson, W. C. Nelson, K. A. Ketchum, L. McDonald, T. R. Utterback, J. A. Malek, K. D. Linher, M. M. Garrett, A. M. Stewart, M. D. Cotton, M. S. Pratt, C. A. Phillips, D. Richardson, J. Heidelberg, G. G. Sutton, R. D. Fleishmann, J. A. Eisen, O. White, S. L. Salzberg, H. O. Smith, J. C. Venter, and C. M. Fraser. 1999. Evidence for lateral gene transfer between Archaea and bacteria from genome sequence of *Thermotoga maritima*. *Nature* **399:**323–329.

Gardner, M. J., H. Tettelin, D. J. Carucci, L. M. Cummings, L. Aravind, E. V. Koonin, S. Shallom, T. Mason, K. Yu, C. Fujii, J. Pederson, K. Shen, J. Hing, C. Aston, A. Lai, D. C. Schwarts, M. Pertea, S. Salzberg, L. Zhou, G. G. Sutton, R. Clayton, O. White, H. O. Smith, C. M. Fraser, M. D. Adams, J. C. Venter, and S. L. Hoffman. 1998. Chromosome 2 sequence of the human malaria parasite *Plasmodium falciparum*. *Science* **282:**1126–1132.

Fraser, C. M., S. J. Norris, G. M. Weinstock, O. White, G. G. Sutton, R. Dodson, M. Gwinn, E. K. Hickey, R. Clayton, K. A. Ketchum, E. Sodergren, J. M. Hardham, M. P. McLeod, S. Salzberg, J. Peterson, H. Khalak, D. Richardson, J. K. Howell, M. Chidambaram, T. Utterback, L. McDonald, P. Artiach, C. Bowman, M. D. Cotton, C. Fujii, S. Garland, B. Hatch, K. Horst, K. Roberts, M. Sandusky, J. Weidman, H. O. Smith, and J. C. Venter. 1998. Complete genome sequence of *Treponema pallidum*, the syphilis spirochete. *Science* **281:**375–388.

Tomb, J. F., O. White, A. R. Kerlavage, R. A. Clayton, G. G. Sutton, R. D. Fleischmann, K. A. Ketchum, H. P. Klenk, S. Gill, B. A. Dougherty, K. Nelson, J. Quackenbush, L. Zhou, E. F. Kirkness, S. Peterson, B. Loftus, D. Richardson, R. Dodson, H. G. Khalak, A. Glodek, K. McKenney, L. M. Fitzgerald, N. Lee, M. D. Adams, E. K. Hickey, D. E. Berg, J. D. Gocyane, T. R. Utterback, J. D. Peterson, J. M. Kelley, M. D. Cotton, J. M. Weidman, C. Fujii, C. Bowman, L. Watthey, E. Wallin, W. S. Hayes, M. Borodovsky, P. D. Karp, H. O. Smith, C. M. Fraser, and J. C. Venter. 1997. The complete genome sequence of the gastric pathogen *Helicobacter pylori*. *Nature* **388:**539–547.

Klenk, H. P., R. A. Clayton, J. F. Tomb, O. White, K. E. Nelson, K. A. Ketchum, R. J. Dodson, M. Gwinn, E. K. Hickey, J. D. Peterson, D. L. Richardson, A. R. Kerlavage, D. E. Graham, N. C. Kyrpides, R. D. Fleischmann, J. Quackenbush, N. H. Lee, G. G. Sutton, S. Gill, E. F.

Kirkness, B. A. Dougherty, K. McKenney, M. D. Adams, B. Loftus, S. Peterson, C. I. Reich, L. K. McNeil, J. H. Badger, A. Glodek, L. Zhou, R. Overbeek, J. D. Gocayne, J. F. Weidman, L. McDonald, T. Utterback, M. D. Cotton, T. Spriggs, P. Artiach, B. P. Kaine, S. M. Sykes, P. W. Sadow, K. P. D'Andrea, C. Bowman, C. Fujii, S. A. Garland, T. M. Mason, G. J. Olsen, C. M. Fraser, H. O. Smith, C. R. Woese, and J. C. Venter. 1997. The complete genome sequence of the hyperthermophilic, sulfate-reducing archaeon *Archaoglobus fulgidus*. *Nature* **390:**364–370.

Fraser, C. M., S. Casjens, W. M. Huang, G. G. Sutton, R. Clayton, R. Lathigra, O. White, K. A. Ketchum, R. Dodson, E. K. Hickey, M. Gwinn, B. Dougherty, J. F. Tomb, R. D. Fleischmann, D. Richardson, J. Peterson, A. R. Kerlavage, J. Quackenbush, S. Salzberg, M. Hanson, R. van Vugt, N. Palmer, M. D. Adams, J. Gocayne, J. Weidman, T. Utterback, L. Watthey, L. McDonald, P. Artiach, C. Bowman, S. Garland, C. Fujii, M. D. Cotton, K. Horst, K. Roberts, B. Hatch, H. O. Smith, and J. C. Venter. 1997. Genomic sequence of a Lyme disease spirochete, *Borrelia burgdorferi. Nature* **390:**580–586.

Fraser, C. M., and R. D. Fleischmann. 1997. Strategies for whole microbial genome sequencing and analysis. *Electrophoresis* **18:**1207–1216.

Bult, C. J., O. White, G. J. Olsen, L. Zhou, R. D. Fleischmann, G. G. Sutton, J. A. Blake, L. M. FitzGerald, R. A. Clayton, J. D. Gocayne, A. R. Kerlavage, B. A. Dougherty, J. F. Tomb, M. D. Adams, C. I. Reich, R. Overbeek, E. F. Kirkness, K. G. Weinstock, J. M. Merrick, A. Glodek, J. L. Scott, N. S. M. Geoghagen, J. F. Weidman, J. L. Fuhrmann, D. Nguyen, T. R. Utterback, J. M. Kelley, J. D. Peterson, P. W. Sadow, M. C. Hanna, M. D. Cotton, K. M. Roberts, M. A. Hurst, B. P. Kaine, M. Borodovsky, H.-P. Klenk, C. M. Fraser, H. O. Smith, C. R. Woese, and J. C. Venter. 1996. Complete genome sequence of the methanogenic archaeon, *Methanococcus jannaschii. Science* **273:**1058–1072.

Fraser, C. M., J. D. Gocayne, O. White, M. D. Adams, R. A. Clayton, R. D. Fleischmann, C. J. Bult, A. R. Kerlavage, G. Sutton, J. M. Kelley, J. L. Fritchman, J. F. Weidman, K. V. Small, M. Sandusky, J. Fuhrmann, D. Nguyen, T. R. Utterback, D. M. Saudek, C. A. Phillips, J. M. Merrick, J.-F. Tomb, B. A. Dougherty, K. F. Bott, P.-C. Hu, T. S. Lucier, S. N. Peterson, H. O. Smith, C. A. Hutchison, III, and J. C. Venter. 1995. The minimal gene complement of *Mycoplasma genitalium*. *Science* **270:**397–403.

Fleischmann, R. D., M. D. Adams, O. White, R. A. Clayton, E. F. Kirkness, A. R. Kerlavage, C. J. Bult, J.-F. Tomb, B. A. Dougherty, J. M. Merrick, K. McKenney, G. Sutton, W. FitzHugh, C. Fields, J. D. Gocayne, J. Scott, R. Shirley, L.-I. Liu, A. Glodek, J. M. Kelley, J. F. Weidman, C. A. Phillips, T. Spriggs, E. Hedblom, M. D. Cotton, T. R. Utterback, M. C. Hanna, D. T. Nguyen, D. M. Saudek, R. C. Brandon, L. D. Fine, J. L. Fritchman, J. L. Fuhrmann, N. S. M. Geoghagen, C. L. Gnehm, L. A. McDonald, K. V. Small, C. M. Fraser, H. O. Smith, and J. C. Venter. 1995. Whole-genome random sequencing and assembly of *Haemophilus influenzae. Rd. Science* **269:**496–512.

EUGENE W. NESTER

Gene Transfer out of the Microbial World

My interest in bacterial genetics was fostered initially in my undergraduate days at Cornell, where I took a course in bacterial genetics under Max Zelle. It was reinforced in my graduate work at Case Western Reserve University where I studied with John Spizizen, the discoverer of DNA-mediated transformation in *Bacillus subtilis*. My graduate studies especially made me realize the power of genetics in analyzing microbial systems, and consequently I chose to strengthen this area by doing postdoctoral work in the small research group of Joshua and Esther Lederberg. They had just left the University of Wisconsin to head a new Department of Genetics at Stanford University School of Medicine.

Because of my graduate background with Spizizen, Josh Lederberg encouraged me to develop DNA-mediated transformation in *Escherichia coli*. His laboratory had already discovered conjugation and transduction, and he thought it would be appropriate that his laboratory close the ring with the discovery of transformation. However, that was not to be. After several months of negative results, we considered the possibility that the donor DNA was being degraded by nucleases. To test this possibility, Charley Yanofsky suggested that we add *Bacillus subtilis* DNA to the cells of *E. coli* before isolating DNA. If nucleases were the problem, then the *B. subtilis* DNA should be degraded and no longer serve as donor DNA. However, the mixture of *E. coli* and *B. subtilis* DNA transformed competent cells of *B. subtilis* very efficiently, and we concluded that it was unlikely that the *E. coli* DNA was the problem. Although this experiment did not provide any insight into what the real problem was, it made us rethink the project and realize that the study of *B. subtilis* transformation might be more fruitful than a continuation of the investigation on *E. coli*. This was an important decision for a postdoc who would be looking for a job in a few years. Josh willingly went along with this decision. His laboratory continued to study *B. subtilis* transformation long after I left Stanford and made many important contributions to the field.

Initially, Josh suggested that we isolate large numbers of mutants and then look for a genetic linkage by transforming double mutants with wild-type DNA. With unusually good luck, we quickly determined that genes of tryptophan synthesis were closely linked to a gene of histidine biosynthesis. More extensive analysis of this region demonstrated that additional genes of aromatic acid biosynthesis were closely linked to each other and to the *his* locus.

With these studies published, I was offered an Assistant Professorship in the Department of Microbiology with a joint appointment in the Department of Genetics at the University of Washington. I recall that when I heard of the opening, I was excited about the university but I knew nothing about Seattle, never having been there. However, I had heard that it was in a lovely part of the country. In Seattle, I continued my studies on the biochemical genetics and regulation of aromatic amino acid biosynthesis, initially in conjunction with two very talented postdocs, Roy Jensen and Delill Nasser. In the course of these studies, we identified a new form of allosteric inhibition of aromatic acid synthesis and published a number of papers on the genetics, biochemistry, and regulation of the pathway. My second area of interest related to the mechanism by which donor DNA is taken up by competent cells of *B. subtilis*. Both of these areas proved valuable in our future studies on plant cell transformation.

After ten years of studying aromatic acid synthesis and competence for DNA transformation in *B. subtilis*, I realized that I had taken these studies about as far as I felt qualified to go. I was not excited by the thought of purifying enzymes and studying their mechanism of action.

Furthermore, graduate students and postdoctoral trainees seemed to be more interested in macromolecule synthesis and gene expression with very minimal interest in small molecule synthesis, and it became difficult to attract good people to the lab. This prompted me to look for new areas of research. I settled on *Agrobacterium* and the disease that it caused in plants, crown gall tumors, for a number of reasons, all of which seem to have crystallized simultaneously. First, President Nixon declared a war on cancer at the national level and indicated that considerable federal funds would be available to study this malady. Crown gall should be included as a plant tumor. Second, Rob Schilperoort of Leiden University in the Netherlands sent his Ph.D. thesis to me in which he reported that nucleic acid of a lysogenic phage in *Agrobacterium* was transferred to plants and caused the tumors. Very exciting stuff! Third, and most important, Milt Gordon, a professor of biochemistry, and Mary-Dell Chilton, a recent Ph.D. in the Department of Genetics, were also anxious to explore new areas of research. Milt had expertise in plant viruses and plant biology, and Mary-Dell was very knowledgeable about nucleic acid technology. Even though none of us had ever taken a course in plant pathology, with my background in microbiology it seemed that we could form a unique team to explore this unusual system of plant cell transformation by a bacterium. We also drew extensively on the expertise of several members of the Botany Department, in particular Arnie Bendich and Walt Halperin.

It was relatively easy to justify this area of research to granting agencies as merely an extension of our previous studies on *B. subtilis* in that tumor formation represented a transformation event that likely involved the transfer of DNA into plant cells. Furthermore, a giant in the field, Dr. Armin Braun at the Rockefeller Institute, had shown that transformed cells could grow in the absence of exogenously added auxin, which untransformed cells require. Auxin is an aromatic compound related to tryptophan. Soon after obtaining some preliminary data, we were successful in getting a grant from the National Institutes of Health. Over the past twenty-five years we have also received support from the National Science Foundation, the American Cancer Society, and the U.S. Department of Agriculture. Postdoctoral fellows have been supported by all of these agencies as well as by the Damon Runyon–Walter Wincell Cancer Research Fund and by the Jane Coffin Childs Foundation. Not many areas of research can claim support from such a wide range of agencies.

From the very beginning, we believed that Schilperoort was correct in concluding that DNA was transferred from *Agrobacterium* into plant cells. However, we also were certain that the DNA of the lysogenic phage that he identified as being transferred into plant cells could not be the "tumor-inducing principle," a phrase that Braun introduced. Using techniques of DNA–DNA hybridization, we demonstrated that many strains of *Agrobacterium* that were not lysogenic for this phage nevertheless caused crown gall tumors. However, since Schilperoort's lab had considerable expertise in the biology of plant tumors, I decided to spend six months on sabbatical leave in his laboratory in the Netherlands, learning about the system and trying to reproduce his experimental data. Rob generously allowed my graduate student, Tom Currier, and me to come and work beside his students. A few months after we arrived, Milt Gordon also came over and spent several months in Rob's lab and several months in Professor George Melcher's lab in Germany, where he produced a number of crown gall cultures from single cells. This was a wonderful experience in my scientific career. Rob was very friendly and helpful, and his students also were always willing to answer all questions. We had many frank and open discussions in an effort to understand the discrepancies between our results and Rob's data.

My experiments in Leiden further convinced me that the phage played no role in tumorigenesis. The addition of deoxyribonuclease to DNA purified from tumor tissue had no effect on its ability to "hybridize" with complementary RNA of the phage. A year or so after I left Leiden, Rob and his colleagues published a paper in which they concluded that a contaminant in their DNA preparations of plant material was responsible for their hybridization results. However, since that time, Rob's laboratory and those of his students, in particular Paul Hooykaas, have made many important contributions to the field of *Agrobacterium* and plant cell transformation.

Therefore, the question of DNA transfer from *Agrobacterium* to plant cells remained unanswered. Before going to the Netherlands, I read a paper in *Experentia* that seemed to hold an important clue to understanding how *Agrobacterium* caused tumors. *Agrobacterium* grows optimally at 28°C. R. Hamilton and M. Fall reported that incubating cells of a particular strain of *Agrobacterium* at 37°C on a solid medium resulted in two sizes of colonies. The larger colonies were avirulent and the smaller colonies remained virulent for tumor formation. Being in the same department as Stanley Falkow and thus aware of the properties and ubiquity of plasmids suggested to us that heat was curing the *Agrobacterium* strain of a plasmid that was required for tumor formation. Therefore, a graduate student in Milt Gordon's laboratory, Bruce Watson, looked intensively for plasmids in the virulent strain. All of his many efforts proved fruitless. However, he was using methods that would only detect small plasmids of the size commonly found in bacteria.

By a stroke of good fortune, while I was working in Schilperoort's laboratory, Rob showed me a preprint of a paper coming out of Jeff Schell's lab in Belgium on which Rob was a coauthor. It presented convincing evidence that all natural isolates of virulent strains of *Agrobacterium* contained a large plasmid, which all avirulent strains lacked. When this preprint, which described the technique for looking for large plasmids, was airmailed back to Seattle, Watson quickly identified a large plasmid in a virulent strain of *Agrobacterium* and showed that the plasmid was absent in the strain made avirulent by heating. As we had surmised, heating had cured the strain of its plasmid, thereby providing convincing proof that the plasmid was required for tumorigenesis. Independently, Jeff Schell and his colleagues, in particular Marc van Montague in Ghent, reached the same conclusion. Thus, a plasmid, which the Belgian group termed the *tumor-inducing* (or Ti) *plasmid*, must be responsible for tumor formation. However, what role it played in tumor formation remained a mystery.

We favored the notion that the plasmid was transferred into plant cells and the expression of plasmid genes provided the new properties to the plant cell. We had already shown, in previous studies carried out before I had gone to Rob's lab, that the entire genome of *Agrobacterium* was not present in tumor tissue. Thus, we now looked for the presence of the entire Ti plasmid in transformed plant cells. To our dismay, these results were also negative. The technique we used was to measure the kinetics of reassociation of small amounts of radioactive denatured plasmid DNA in the presence of large amounts of denatured tumor or normal tissue DNA. If plasmid sequences were present in the tumor tissue, then the increased concentration of the plasmid sequences would result in a more rapid renaturation of the radioactive plasmid DNA. However, this technique is not sensitive enough to detect the presence of only a few genes of the large Ti plasmid.

Therefore, we next considered the possibility that only a small piece of the Ti plasmid was transferred. We had extensive discussions with

A crown gall tumor induced by Agrobacterium tumefaciens *on a* datura *plant.*

our colleagues and among ourselves about whether these experiments were worth undertaking. Several of our colleagues argued that the transfer of DNA from a prokaryote to a eukaryote and its stable incorporation into the eukaryotic cell were very far fetched. We were well aware from our previous studies on DNA transformation in *B. subtilis* how difficult it was to achieve any degree of stable DNA transfer when even closely related bacteria were the source of donor DNA. However, all data pointed to transfer of DNA, so we did not give up the hunt.

The challenge was to identify this putative fragment of the plasmid present in tumor tissue.

Although studies to demonstrate DNA transfer and incorporation are straightforward and relatively simple now, in the mid-1970s they presented several challenges. Not many restriction enzymes were commercially available. Cloning was more an art than a science, and the isotopic labeling of DNA to high specific activities was not an established and reproducible procedure. Southern blotting technology was in its infancy. We used one of the few restriction enzymes available, Sma 1, and the same technique of reassociation kinetics that we had used to look for the entire Ti plasmid in tumor tissue. This enzyme cleaved the Ti plasmid into seventeen fragments. We then looked for each fragment in transformed tissue. Using small fragments of DNA greatly increased the sensitivity of the assay. As a control, we did the same experiment using untransformed tissue. To our great excitement and sense of relief, one restriction fragment gave consistently positive results. After we had repeated this "brute force" experiment several times, taking out many samples at various times over a several-day period and determining the kinetics of reassociation of the DNA, we were convinced that *Agrobacterium* does indeed have the unique ability to transfer a small piece of its plasmid into plant cells. We termed this DNA T-DNA for tranferred DNA.

Coming quickly on the heels of this observation, our laboratory demonstrated that the plasmid genes were transcribed into RNA and that the plasmid DNA was covalently integrated into the plant genome. Furthermore, our laboratory and others demonstrated that the transferred or T-DNA coded for genes of auxin and cytokinin biosynthesis. We and others also demonstrated that another set of genes that we termed the virulence (*vir*) genes, also located on the Ti plasmid, were required for the transfer of the T-DNA into plant cells. To our surprise, we also found that some of the oncogenes of the Ti plasmid are present in normal tobacco plants.

Numerous laboratories in the United States and abroad continue to actively study the biology of *Agrobacterium* and its association with plants. These studies proceed along two major paths. One relates to the basic biology of the association and the molecular mechanism of plant cell transformation. The other focuses on attempts to develop *Agrobacterium* into a more efficient vector for the genetic engineering of all higher plants. As far as we know, *Agrobacterium* is unique in its ability to transfer DNA into eukaryotic cells. However, it is now clear that many plant and animal pathogens can transfer proteins into their host cells. What is especially intriguing is that many animal pathogens, such as *Bordetella*, *Legionella*, *Brucella*, and *Helicobacter*, have genes similar in nucleotide sequence to the genes, which *Agrobacterium* uses to transfer T-DNA. At least some of these genes have been shown to be essential for the organisms to cause disease. Thus, *Agrobacterium* serves as a model for protein transfer in many animal pathogens.

Agrobacterium is the workhorse for generating transgenic plants in plant biotechnology. Experiments have shown that *Agrobacterium* can transform most of the economically important grasses such as rice, wheat, and maize. However, other recalcitrant plants still resist efficient transformation and subsequent regeneration into plants. Accordingly, experiments are in progress in many laboratories to make *Agrobacterium* a more efficient vector for transforming all plants.

The role that *Agrobacterium* has played and continues to play in plant biotechnology and in plant molecular biology and genetics represents an excellent example of the unforeseen ramifications of basic research. We initiated our studies with the goal to understand how a common soil bacterium could stably alter the properties of the plants it infects. We could not possibly have foreseen at that time the significance of this system. Other investigators have taken our early results and developed this system into the cornerstone of plant biotechnology. Furthermore, because of the ability of *Agrobacterium* to transfer desired genes between plants, enormous advances have been made in plant molecular biology and developmental genetics.

One of the most gratifying and stimulating aspects of working in this area of plant cell transformation has been the great number of talented, motivated, and hardworking individuals who were attracted to our laboratory, either as postdoctoral trainees, graduate students, or visiting scientists. Most have gone on to develop independent careers, and many have continued to study and make important contributions to our knowledge about *Agrobacterium*. I am eternally grateful to Milt Gordon and Mary-Dell Chilton and all who passed through my laboratory for making the studies of *Agrobacterium* such an exhilarating and rewarding experience. This essay is dedicated to all of them, without whose efforts this story could not have been written.

EUGENE W. NESTER received his bachelor of science degree from Cornell University in 1952 and his doctoral degree from Case Western Reserve University in 1959. He joined the faculty of the University of Washington in 1962. He is currently a professor in the Department of Microbiology at the University of Washington. He is a member of the National Academy of Sciences and has received numerous awards, including the Cetus Biotechnology Research Award and the inaugural Australia Prize. In 1999 he was elected chair of the Board of Governors of the American Academy of Microbiology.

The following papers are representative of his publications:

Stephens, K. M., C. Roush, and E. Nester. 1995. *Agrobacterium tumefaciens* VirB11 protein requires a consensus nucleotide-binding site for function in virulence. *J. Bacteriol.* **177:**27–36.

Toro, N., A. Datta, M. Yanofsky, and E. Nester. 1988. Role of the overdrive sequence in T-DNA border cleavage in *Agrobacterium*. *Proc. Natl. Acad. Sci. USA* **85:**8558–8562.

Das, A., S. Stachel, P. Ebert, P. Allenza, A. Montoya, and E. Nester. 1986. Promoters of *Agrobacterium tumefaciens* Ti-plasmid virulence genes. *Nucleic Acids Res.* **14:**1355–1364.

Yanofsky, M., A. Montoya, V. Knauf, B. Lowe, M. Gordon, and E. Nester. 1985. Limited-host-range plasmid of *Agrobacterium tumefaciens*: Molecular and genetic analyses of transferred DNA. *J. Bacteriol.* **163:**341–348.

Douglas, C., W. Halperin, M. Gordon, and E. Nester. 1985. Specific attachment of *Agrobacterium tumefaciens* to bamboo cells in suspension cultures. *J. Bacteriol.* **161:**764–766.

Knauf, V., M. Yanofsky, A. Montoya, E. Nester. 1984. Physical and functional map of an *Agrobacterium tumefaciens* tumor-inducing plasmid that confers a narrow host range. *J. Bacteriol.* **160:**564–568.

Lichtenstein, C., H. Klee, A. Montoya, D. Garfinkel, S. Fuller, C. Flores, E. Nester, and M. Gordon. 1984. Nucleotide sequence and transcript mapping of the *tmr* gene of the pTiA6NC octopine Ti-plasmid: a bacterial gene involved in plant tumorigenesis. *J. Molec. Appl. Gen.* **2:**354–362.

Klee, H., A. Montoya, F. Horodyski, C. Lichtenstein, D. Garfinkel, S. Fuller, C. Flores, J. Peschon, E. Nester, and M. Gordon. 1984. Nucleotide sequence of the *tms* genes of the pTiA6NC octopine Ti plasmid: two gene products involved in plant tumorigenesis. *Proc. Natl. Acad. Sci. USA* **81:**1728–1732.

Huang, L., M. Nakatsukasa, and E. Nester. 1974. Regulation of aromatic amino acid biosynthesis in *Bacillus subtilis* 168. Purification, characterization, and subunit structure of the bifunctional enzyme 3-deoxy-D-arabinoheptulosonate 7-phosphate synthetase-chorismate mutase. *J. Biol. Chem.* **249:**4467–4472.

Listening to What the "Bug" Tells You

Like many students going through the public school system in New York in the 1940s, I was exposed to the standard array of science courses—chemistry, physics, and biology—three separate and unrelated learning experiences. The College of Agriculture at Cornell University was virtually cost-free to residents of New York State and so I decided to go there. But studying biology was complicated because some courses were taught in the "Ag" school and others in Arts and Sciences. As an Aggie, I was required to take farm practice courses, which were obtained by working summers on dairy, fruit, vegetable, and other farms. I worked my way through a myriad of courses from pomology to animal nutrition to microbiology and beyond. In a course taught by Adrian Srb, I was exposed for the first time to genetics, particularly microbial genetics. I knew for certain that a career in microbiology and microbial genetics was my destiny. Looking back, that interest was rooted in my childhood, when I bred racing pigeons on a rooftop in New York.

Although there were laboratories in zoology, botany, and geology (and I took them all), the microbiology laboratories were "user friendly." Perhaps what was so exciting was the hands-on approach.

Results were in real time and experiments could be performed, even when they were not part of the formal laboratory exercise. This was certainly not true in other areas. During Srb's course we discussed Euphrusi's early work with *Drosophila* and McClintock's work with corn. However, it was the *Neurospora* genetics of Beadle and Tatum—to which Euphrusi had contributed—and Euphrusi's work with yeast and the beginnings of *Escherichia coli* genetics or bacteriophage T4 that seemed more meaningful. I found a seminar by David Bonner—Srb's Ph.D. mentor—particularly exciting. It described immunologic cross-reacting material in mutants of *Neurospora crassa*. Despite losing the ability to make tryptophan, the mutants still produced a protein immunologically related to the native enzyme, tryptophan synthetase. From Srb's course, the work of Beadle, Tatum, Euphrusi, Benzer, and others promoting the one-gene, one-enzyme hypothesis could be modified; a simple point mutation could result in a single amino acid change in a protein without causing the loss of the entire protein. Although taken for granted today, this finding was profound at the time.

Without detracting from the early genetic work, the developing paradigm achievable through the study of microorganisms was a far cry from changes in eye color or wing veination in *Drosophila* and starch accumulation of corn kernels. The microbial experimental system allowed one to get up close and friendly with the research and to ultimately consider the question of mechanism. David Bonner's seminar was enough to convince me that I wanted to work toward a Ph.D. and to work with him. Until then, I had never considered an academic research career in microbiology. I was clearly at the right place at the right time.

I spent two years at Yale University with Dave Bonner as his Ph.D. student. When he went to the University of California, San Diego (UCSD), to start the new Department of Biology, I went with him. I was the first student to earn a Ph.D. in biology at UCSD. Several years ago, I had to prove I had received a Ph.D. To my chagrin, UCSD claimed there was no record of my having ever received a degree there. After a struggle with the registrar's office and numerous phone calls spent listening to elevator music while I was placed on hold, I was referred to the "Archives," where lo and behold, the record of my degree was entombed. Interestingly, the Alumni Association knows precisely who I am and where to reach me at all times to ask for money.

While at Yale, I took a course in microbial physiology, with lectures by Doudoroff, Wood, DeMoss, and other exceptional microbiologists. Lectures by the late Wolfe Vishniac dealt with "funny bugs," an anachronism describing a collection of bacteria that were then far outside the mainstream—for example, methane, sulfur, and photosynthetic bacteria. These bacteria and archaea have now joined the mainstream, and yesterday's "funny bugs" are today's models. So keep an open mind!

The photosynthetic bacteria were particularly interesting to me because I was amazed that nonsulfur purple bacteria could grow heterotrophically and photosynthetically or not, depending on the absence or presence of oxygen, switching metabolic modes based on levels of oxygen. I realized the photosynthetic membrane system of the purple nonsulfur bacteria could be useful for investigating genetic control of membrane structure, function, and synthesis. This may have been my first original scientific thought. Again, I was in the right place at the right time.

During the remainder of my Ph.D. training and my postdoctoral training with Sydney Brenner, working on amber suppressors in bacteriophage T4, I followed the literature pertaining to genetic control of photosynthetic membrane development in the facultative photoheterotroph *Rhodobacter sphaeroides*. My first academic position was in the Department of Cell Biology at Western Reserve University in Cleveland, Ohio, in 1967. I remained there for only six months, as I was caught in the middle of a power struggle between various elements of the university—my first exposure to university politics, but not my last.

I moved to the University of Illinois at Champaign-Urbana. My first grant application, entitled "Induction and Biogenesis of Subcellular Organelles," was funded by the National Institutes of Health and continues to be funded in its thirty-seventh year. Examination today of the original proposal reveals two things: how far we have come and how little we really know. Barry Marrs, who had been a student of Boris Euphrusi's wife, Harriett Euphrusi-Taylor, became my postdoctoral researcher. His training in my laboratory was his introduction to the photosynthetic bacteria. Subsequently, he went on to study *Rhodobacter capsulatus*. Between his laboratory and mine, we truly developed the molecular approaches to the study of the photosynthetic bacteria.

To investigate the photosynthetic membranes of the purple nonsulfur bacteria, it was essential to develop a genetic system for *R. sphaeroides*. Such a system could be applied to study the control of gene expression by oxygen and light. However, I was not so foolish as to embark on the development of a new experimental system with which I had no prior experience without maintaining some links to the past, just in case the system proved intractable or I proved inadequate for the task. Thus, during the transition from postdoctoral to independent investigator at Illinois, our laboratory continued working with *Escherichia coli*. Graduate and postdoctoral students made innumerable findings on the biochemistry and genetics of ribosomal RNA gene regulation using *E. coli* as a model. Eventually we had the tools and findings needed to get out of the *E. coli* business and concentrate on the inducible membrane system of *R. sphaeroides*. With the advent of recombinant DNA technology in 1974, our laboratory became more successful in the pursuit of our goals, studying the genetics, regulation, and biosynthesis of the photosynthetic membranes of *R. sphaeroides*. As so often happens, each new avenue explored yielded numerous unintended surprises, and microorganisms are full of surprises!

But like most laboratories, the incorporation and exploitation of new technologies rest squarely in the hands of new graduate students and postdoctorals. Chance continued to smile and I was fortunate to become associated with some truly remarkable graduate students. Throughout the 1980s things were really booming. Our laboratory was virtually running around the clock. Students and postdoctorals were there day and night. Why not? What else was there to do in Champaign-Urbana, Illinois?

In 1989 I decided to move the laboratory from Illinois to Houston. The move was timely: My children had left for college, and I had just spent a year in Geneva, Switzerland, enjoying the irresistible delights of the big city. Academic politics at Illinois had made my continued presence unwelcome, and I could start a new Department of Microbiology. Moreover, the research was poised to move to a new level. Antonius Suwanto, a graduate student who began his studies at Illinois constructing a physical map of the *R. sphaeroides* 2.4.1 genome using the very new technique of pulsed-field electrophoresis, discovered that *R. sphaeroides* possesses two unique chromosomes. This discovery was not immediately accepted, even in our own laboratory. I was among those who found it difficult to accept. Dogma was that bacteria had a single chromosome. The axiom that has always existed in our laboratory—listen to what the "bug" is trying to tell you—had been forgotten.

Subsequently, bacteria have been shown to contain linear chromosomes, and at least six additional groups of bacteria (and the numbers continue to increase) have been demonstrated to have more than a single chromosome. Thus, larger and more significant questions loom about why multiple chromosomes occur in bacteria: What is the evolutionary relationship(s) between those organisms containing multiple chromosomes to one another and to those organisms with a single chromosome? It is remarkable that in our investigation of multiple chromosomes in *R. sphaeroides*, our laboratory should return to the *trp* operon. Our studies have shown that the genes encoding the enzymes for the biosynthesis of L-tryptophan are located on each of the chromosomes and that their relative position and topologies are unusual. Just when you are convinced that you are beginning to gain a fundamental understanding of questions posed at an earlier time, you serendipitously unveil yet another fundamental insight. It is a lot like peeling an onion, but with microorganisms the revelations are so immediate and so intense, the wonder never ceases.

The other major new initiative dating to 1989 derived from the work of graduate student Jeong Lee (also an Illinois transplant), who developed genetic approaches to defining several of the major regulatory elements responding to O_2 and light. We revealed how the interaction of living systems with O_2 and light can lead to signaling, through specific redox chains and carriers, to unique regulatory proteins that serve to regulate gene expression. This work is clearly the first example of how a specific redox chain can "talk" to a diverse array of downstream genes and serves as a model for "redox signaling" in both lower and higher organisms. These studies have also revealed how the cellular flow of reducing equivalents can be used to control both transcriptional and posttranscriptional cellular processes.

As I have already said, the removal of each carefully structured layer unveils surprises. My personal joy from my experience as a microbiologist comes from my participation in the process of discovery. For an instant, no matter how brief, to have learned something that no one else has apparently ever known, whether it be processing ribosomal RNA, multiple chromosomes, or redox signaling, is an immensely satisfying feeling. To be challenged almost daily by these remarkable creatures and to devise experimental protocols and questions that enable these complex organisms to reveal their secrets garnered over a billion years of evolutionary time are on the one hand humbling and on the other hand satisfying beyond belief. Experimentation is really the art of getting these creatures to reveal themselves to you. Discovery, challenge, understanding, imagination, and appreciation are all key words in describing the microbiologist.

Of equal satisfaction is sharing these goals with students and colleagues in an effort to unlock the secrets of life for the good of humankind. Discovery in a vacuum is nonexistent, but to see a student awaken as a result of the process of discovery adds immeasurable satisfaction to the event. Certainly, such satisfaction can occur in any field, science or otherwise, but because of their immense biochemical versatility, ubiquity, numbers, and ease of experimental manipulation, microorganisms deliver this satisfaction in real time and with considerable impact. When you couple the immediacy of the experimental interaction when working with microorganisms to the sharing of such experiences with students and postdoctorals, the impact and reward are profound. Teaching undergraduate students—something I have done for over twenty years—is equally rewarding. Undergraduates enjoy hearing of yesterday's experiments and learning about what we don't know. To teach a science as a series of known facts, principles, or concepts is to fossilize the discipline.

SAMUEL KAPLAN was born in Yonkers, New York, in 1934. He received his bachelor of science degree from Cornell, his master's degree from Yale, and his doctoral degree from the University of California, San Diego. He has served on the faculty of the University of Illinois, Chămpaign-Urbana. He is head of the Publications Board of the American Society for Microbiology and is currently a professor and chair of Microbiology and Molecular Genetics at the University of Texas Medical School at Houston. One of his aims is to train students so thoroughly that they understand what quality science means and are able to conduct their own independent research program in any field, even if they never did a postdoctoral fellowship.

The following papers are representative of his publications:

O'Gara, J., and S. Kaplan. 1997. Evidence of the role of redox carriers in photosynthesis gene expression and carotenoid biosynthesis in *Rhodobacter sphaeroides* 2.4.1. *J. Bacteriol.* **179:**1951–1961.

Suwanto, A., and S. Kaplan. 1989. Physical and genetic mapping of the *Rhodobacter sphaeroides* 2.4.1 genome: the presence of two unique circular chromosomes. *J. Bacteriol.* **171:**5850–5859.

Kiley, P. J., A. Varga, and S. Kaplan. 1988. A physiological and structural analysis of light harvesting mutants of *Rhodobacter sphaeroides*. *J. Bacteriol.* **170:**1103–1115.

Zhu, Y. S., and S. Kaplan. 1985. The effects of light, oxygen, and substrates on steady-state levels of mRNA coding for ribulose-1,5-bisphosphate carboxylase, light harvesting, and reacting center polypeptides in *Rhodopseudomonas sphaeroides*. *J. Bacteriol.* **162:**925–932.

Chory, J., T. J. Donohue, A. R. Varga, L. A. Staehelin, and S. Kaplan. 1984. Induction of the photosynthetic membrane of *Rhodopseudomonas sphaeroides*: biochemical and morphological studies. *J. Bacteriol.* **159:**540–554.

Fraley, R. T., C. S. Fornari, and S. Kaplan. 1979. Entrapment of a bacterial plasmid in phospholipid vesicles: potential for gene transfer. *Proc. Natl. Acad. Sci. USA* **76:**3348–3352.

Kaplan, S., A. Atherly, and A. Barrett. 1973. Synthesis of stable RNA in stringent *Escherichia coli* cells in the absence of charged tRNA. *Proc. Natl. Acad. Sci. USA* **70:**689–692.

Marrs, B., and S. Kaplan. 1970. 23S precursor ribosomal RNA of *Rhodopseudomonas sphaeroides*. *J. Mol. Biol.* **49:**297–317.

Kaplan, S., A. O. W. Stretton, and S. Brenner. 1965. Amber suppression: Efficiency of chain propagation and suppressor specific amino acids. *J. Mol. Biol.* **14:**528–533.

Kaplan, S., Y. Suyama, and D. M. Bonner. 1964. Fine structure analysis at the Td locus of *Neurospora crassa*. *Genetics* **49:**145–158.

MOSELIO SCHAECHTER

Why I Am Amazed by Simple Things

As is true for all researchers, a new result in the lab produces a surge of excitement and renewed enthusiasm. It is hard to describe the "high" that accompanies even a modest discovery. Sometimes this comes about in a simple way, such as by looking at colonies on a Petri dish or at an x-ray film of an electrophoresis gel and realizing that here is an answer to a question that had not been asked before. Other times, the breakthrough follows the use of complex technology. I reserve my sense of primordial awe for one of the simplest experiments in bacterial physiology that is done over and over in laboratories. It is the measurement of the growth of a bacterial culture in a liquid medium. One can get an instantaneous reading simply by determining the turbidity of the culture at different times using a common light-measuring device such as a colorimeter. With time, as the bacteria grow, the turbidity increases and the readings increase. Of course, what is so spectacular is that in a rich medium, the bacterial mass doubles in twenty minutes or less. And it does it every time, like clockwork. I find it hard to imagine how everything that goes into making bacteria—their enzymes, genes, structural elements—doubles so precisely and so rapidly. No wonder that when I feel down, I go to the bench and "run a growth curve"!

In our age, most experiments depend on sophisticated technology—appropriately so. Let me nevertheless try to explain, using my experience from a simpler age, why I still get excited about running growth curves. Until the mid-1950s, the growth of bacteria was something of a mystical subject. Cultures were known to go through stages: a lag phase, an exponential phase of rapid growth, and a stationary phase. (As a pedantic aside, the exponential phase is commonly called the "log phase," which is OK as lab parlance but is mathematically incorrect.) Drawn on logarithmic paper, the curve looks S-shaped, which, in the old days, invited much theorizing about its deeper meaning. Many models were proposed based on the idea that these various phases of growth were inevitable, and that a bacteria culture had to undergo all of them for some preordained purpose. This, it turns out, was the wrong way to look at it. I was involved in the dispelling of this myth and in attempting to clarify what really goes on when bacteria grow in the laboratory.

In the 1950s, I was working in the laboratory of a distinguished Danish microbiologist, Ole Maaløe. In addition to enjoying the delights of Copenhagen, I became involved in research on bacterial growth physiology using the enteric bacterium *Salmonella typhimurium*. We based our work on the known fact that a given species of bacteria will grow more rapidly in a nutritionally complex medium, a so-called rich medium, than in a "minimal medium" in which the only organic substance may be a simple sugar such as glucose. The question we asked was: What happens when the cells are placed abruptly in a different medium? How do they adjust to the new conditions? How long do they take to make this adjustment? One of the experiments we did was to add concentrated rich medium to a culture in a poor medium and to follow the increase in turbidity as well as in cell number. We learned that after such a nutritional "shift-up," the cellular mass increases at the new rate immediately, with a nearly imperceptible lag. The increase in the number of cells, on the other hand, did not proceed at the new rate until quite some time later. Why the discrepancy between the increase in mass (which is what turbidity measures) and cell number? The simplest explanation was that cells growing in the poor medium were smaller than those growing in the rich medium. After the shift from poor to rich, the cells had to become larger. Thus, they grew in mass but for a while did not divide, which explains the delay in the rate of increase in numbers.

Why would cells of the same species differ in average size? We wondered if this was an intrinsic property, dependent on the rate of growth alone, rather than on the composition of the medium. We set up cultures in a collection of different media that supported various growth rates, from the slowest to the fastest attainable in that laboratory. The range of generation times was from two hours to seventeen minutes per doubling at the optimum temperature, 37°C. What we found is that indeed there was a simple relationship between mass and rate of growth, regardless of the *composition of the medium*. In other words, cells growing in two different media but at the same growth rate have the same cell size.

This finding, though removing some of the mystery surrounding bacterial growth, led to the next question. What is it about the size of cells that is influenced by the growth rate? How should one think about it? Bacteria consist mainly of proteins, which account for half or more of their dry weight. We wondered if fast-growing cells weren't larger because they contained more ribosomes, the protein-synthesizing apparatus. We measured the content of ribosomes in cells growing at different rates and found, to our delight, that there was a simple relationship here too. The faster the growth rate, the more ribosomes per cell mass. In other words, the concentration of ribosomes turned out to be a linear function of the growth rate. As if to test the rule, this relationship breaks down at very slow rates, which makes sense because otherwise cells growing infinitely slowly would have no ribosomes! Such cells would not be able to make proteins when placed in a better medium.

What does this linear relationship between ribosome content and growth rate tell us? First, the rate of protein synthesis is proportional to the growth rate, as long as cells grow unhindered. This means that the *concentration* of ribosomes is proportional to the *rate* of protein synthesis. In faster growth, the need to make more protein per minute is met by making more ribosomes, and conversely. In turn, this tells us that ribosomes operate at a single unit of efficiency, regardless of whether they find themselves in a small, slow-growing cell or a large, fast-growing one. Another way of expressing this is that the rate of polymerization of proteins (the chain growth rate) is *constant* as long as cells are growing under what is known as balanced growth conditions. The chain growth rate was eventually measured directly by others, and it turns out that at 37°C, bacteria hook together, on average, fourteen amino acids per second regardless of the medium. In time, it was found that the concept that the polymerizing machinery of bacteria performs at unit rates is also true for the biosynthesis of DNA, RNA, and cell wall constituents. This finding demonstrates the economy that bacteria exhibit in adapting to different growth environments. Instead of making the same amount of biosynthetic machinery under all conditions, which would be a burdensome and wasteful strain on their economy, they make only what they actually need in a given condition. Thus, bacteria in different media are different. They obey the maxim of the Spanish philosopher Ortega y Gasset: "I am I and my circumstance" (*Yo soy yo y mi circunstancia*).

This way of thinking has helped others in further experimentation that revealed a great deal about the mechanisms that control gene expression in bacteria. How is the synthesis of the RNA of the ribosomes regulated? What about the synthesis of ribosomal proteins? What does this have to do with how gene expression is regulated? Questions of this sort deal with the central problems of biological regulation, and much has been learned from sophisticated and elaborate experiments that take advantage of a combination of physiological thinking and genetic tools. I have participated in this work and derive much pleasure from the sophisticated understanding of the mechanisms that have been unraveled. However, I still stand in awe at the central marvel, the ability of such seemingly simple cells to grow in such perfect rhythm.

MOSELIO SCHAECHTER was born in Milan, Italy, in 1928, and was raised in Quito, Ecuador. He received degrees from the Universities of Kansas and Pennsylvania. He chaired the Department of Microbiology and Molecular Biology at Tufts University in Boston, where he is Emeritus Distinguished Professor of Microbiology. He was president of the American Society of Microbiology from 1985 to 1986. After his initial work on growth physiology of bacteria, he studied the role of the cell membrane in bacterial DNA replication and segregation.

The following books and papers are representative of his many publications:

Schaechter, M., G. Medoff, and B. Eisenstein. 1998. *Mechanisms of Microbial Disease,* 3rd ed. Williams & Wilkins, Baltimore.

Schaechter, E. 1997. *In the Company of Mushrooms. A Biologist's Tale.* Harvard University Press, Cambridge, Mass.

Neidhardt, F. C., R. Curtiss, III, J. L. Ingraham, E. C. C. Lin, K. B. Low, B. Magasanik, W. S. Reznikoff, M. Riley, M. Schaechter, and H. E. Umbarger (ed.). 1996. *Escherichia coli and Salmonella*: *Cellular and Molecular Biology,* 2nd ed. ASM Press, Washington, D.C.

Neidhardt, F. C., J. L. Ingraham, and M. Schaechter. 1990. *Physiology of the Bacterial Cell.* Sinauer Associates, Sunderland, Mass.

Schaechter, M., F. C. Neidhardt, J. Ingraham, and N. O. Kjeldgaard (ed.). 1985. *The Molecular Biology of Bacterial Growth.* Jones & Bartlett, Boston, Mass.

Abe, M., C. Brown, W. Henderickson, D. Boyd, P. Clifford, R. Cote, and M. Schaechter. 1977. The release of *E. coli* DNA from membrane complexes by single strand endonucleases. *Proc. Natl. Acad. Sci. USA* **74:**2756–2760.

Green, E. W., and M. Schaechter. 1972. The mode of segregation of the bacterial cell membrane. *Proc. Natl. Acad. Sci. USA* **69:**2312–2316.

Schaechter, M. 1963. Bacterial polyribosomes and their participation in protein synthesis *in vitro. J. Molec. Biol.* **7:**561–568.

Koch, A. L., and M. Schaechter. 1962. A model for statistics of the cell division process. *J. Gen. Microbiol.* **29:**435–454.

Schaechter, M., O. Maaløe, and N. O. Kjeldgaard. 1958. Dependency on medium and temperature of cell size and chemical composition during balanced growth of *Salmonella typhimurium. J. Gen. Microbiol.* **19:**592.

Section Eight

The Biotechnological Revolution

"But what we want, what the enterprising city of Lille wants most of all, professor, you can hear the Committee of business men telling him, is a close cooperation between your science and our industries. What we want to know is—does science pay?" For Pasteur the answer was simple—there was no distinction between basic and applied research. "Where in your families will you find a young man whose curiosity and interest will not immediately be awakened when you put into his hands a potato, and when with that potato he may produce sugar, and with that sugar alcohol, and with that alcohol ether and vinegar?" And so industrial microbiology was both the benefactor and beneficiary of the science of microbiology.

But it was the unraveling of the genetic code and the subsequent discovery of the means to cut, splice, and clone genes that brought with it the power to cross species barriers—to change living organisms. And with that came the power to fuel the biotechnological revolution—and to change science and society. Bacteria could be made to produce human insulin and human growth factor. Sheep could be cloned. Humans could be changed, cured of dreadful inherited diseases by gene therapy. Scientists could design life. Scientists became entrepreneurs, and investment bankers backed the small biotechnology start-up companies that would grow, merge, and prosper.

But with the biotechnological revolution came conflict. Activists protested the dangers of the power of genetic engineering. The scientific community responded with the Asilomar conference, the ensuing moratorium on recombinant DNA technology, and eventually the Recombinant DNA Advisory Committee of the National Institutes of Health to reassure a worried public. And while quelling public concern, the scientific community itself was being tested by the promises of riches that were to be made. Patents replaced peer-reviewed publications. Battle lines were drawn over intellectual property. Who would own the microbes? Who could own life?

> The university is a world of scholarship and trust, where the reward for success is intellectual recognition. Industry is a world of contracts and insecurity, where pay is the reward for work, and success may make one expendable,"
>
> —Salvador Luria, *Slot Machines, A Broken Test Tube: An Autobiography*

And all the time stocks were traded. Money was to be made through biotechnology. And the markets rose.

28 JEAN BRENCHLEY

Putting Microorganisms to Work

My love of microorganisms started while growing up on a small dairy farm in Pennsylvania. The thought that unseen life was all around me stirred my imagination. Because I was unable to watch microorganisms the way I watched cows grazing in a pasture, I had to imagine the way they lived. This became a game, especially when I learned that we actually put these invisible organisms to work making ensilage for the cows and pickles and sauerkraut for ourselves. Later in college, while my friends gazed at the night sky awestruck at "billions and billions" of stars, I marveled at the billions and billions of microbes beneath my feet—eating, breathing, and growing. I relished the idea that, through colony morphologies and counts, I might someday be able to understand how these clever microbes flourished.

Turning that love and curiosity into a career, however, was neither straightforward nor easy. I received my bachelor's degree in 1962, a time when professors discouraged women from science careers and graduate school. Despite my top grades and almost full-time work as a lab assistant, whenever I inquired about advanced study, my professors suggested my proper place was teaching at a nearby high school.

Finally, after applying with great determination to many graduate schools, I was admitted to the Scripps Institute of Oceanography of the University of California, San Diego (UCSD). There, too, women were treated somewhat as unwelcome intruders and aliens, but I did get my first glimpse of the fascinating new sciences of molecular biology and microbial genetics. My first immersion in microbiology came when I seized the chance to attend the University of California, Berkeley, for a semester in an inter-campus exchange. There I fell forever under the spell of microbiology, watching the huge figure of Roger Stanier stride across the lecture hall, revealing wonder after wonder of the microbial world, and Michael Doudoroff and George Hegeman, like a pair of detectives searching for some new microbe, delighting in the revelation of their latest scheme to enrich for an anaerobic, nitrogen-fixing, cellulose-degrading spore-former.

This experience whetted my appetite for more microbiology, and after receiving a master's degree from UCSD, I found a home in Dr. John Ingraham's laboratory at the University of California, Davis. There, I dove into mutant hunts and genetic crosses. The biochemistry of the cell was a big picture puzzle. The emerging field of microbial genetics suddenly allowed thinking of the puzzle as pieces and permitted working on them one piece at a time. Instead of having to observe the chaos of a thousand simultaneous reactions, we could select mutants with one activity eliminated and examine the consequence of that specific loss to the growth of the cell. The affected gene could then be mapped relative to others using genetic crosses. Once thousands of mutations were mapped, a splendid picture of the organism's chromosomal arrangement emerged.

Continuing to apply the new power of genetics, I went on to study nitrogen metabolism with Boris Magasanik during my postdoctoral work at MIT. By isolating the first glutamine and glutamate auxotrophs of *Klebsiella aerogenes*, we discovered that the glutamate dehydrogenase activity could be eliminated without major consequence to cell growth. This was a surprising finding because the highly characterized glutamate dehydrogenase was thought to be the main enzyme for glutamate formation. We showed that an additional mutation was needed to inactivate a second enzyme, glutamate synthase, to create a glutamate requirement. The search to understand the convoluted regulation of nitrogen assimilation into glutamate and glutamine continued into my faculty days at Purdue University in the late 1970s. By then, the new breakthrough of recombinant DNA technology lifted our uses of microbial genetics to a whole new level. Now we could look at genes not just as locations on a chromosomal map but also as actual nucleotide sequences determined from cloned genes.

Recombinant DNA methods changed more than our ability to unravel the mysteries of microbial genetics. They revolutionized the ways we could put microorganisms to work for us. On the farm I saw how microbes gave us foods, and in college I learned how *Penicillium* and *Streptomyces* gave us the gifts of lifesaving antibiotics. Now genetic engineering enabled scientists to teach microorganisms new and better tricks. They could isolate DNA not only from microorganisms but also from eukaryotic plant and animal cells as well. The purified DNA could be recombined in vitro with a plasmid and transformed into a bacterium, such as *Escherichia coli*, turning the host into a miniature factory for making novel products. Newspaper headlines boldly promised marvelous uses ranging from making human insulin to creating nitrogen-fixing plants. The proclamations carried by the lay media fueled the excitement and attracted investment money to industrial microbiology, or biotechnology, as this new thrust of applying recombinant DNA methods to make commercially useful products became known. In the early 1980s, scientists and business organizers were rapidly joining forces to form hundreds of new biotechnology companies.

At Purdue, newly formed companies actively recruited my students and me. At first, I declined to take time from my research to interview with these fledgling enterprises. But then my students began asking me, "Should we consider jobs with them? Please visit these places and size them up for us." The visits transformed my career and view of science.

In my research at Purdue, our goal was to understand how the enzymes making glutamate and glutamine were regulated. We had no reason to ask how much amino acid was produced. But at one of the companies I visited—one so new it consisted only of offices above a restaurant—I realized that increasing the amount of a product was exactly what biotechnology companies sought most. I further discovered that we could apply the same approaches we used to understand regulatory mechanisms to increase the amount of a product. As I talked with these new teams of scientists and entrepreneurs, my ideas flowed and my excitement heightened. Suddenly I wanted to be part of this grand new endeavor.

I accepted an offer to become a research director at a small biotechnology company that had just doubled in size to about eighty employees. Although the official job was to oversee scientific projects, one of the joys of being with a small company on the move was that directors were involved in everything. We met with clients, pitched our successes to investors (an exciting form of teaching), fought to keep projects moving, and invented new products. The action was nonstop and the company's rapid growth—while always on the perilous edge of bankruptcy—kept us on an exhilarating roller coaster ride. It was the most rapid learning experience I had known since graduate school.

Life in a biotechnology company also gave me a broader view of science. In basic research, we had looked at one problem in great depth. In the biotechnology world, we looked at a landscape of related research problems. Instead of pursuing one pathway in one organism, we tracked a complex array of pathways, using *Corynebacterium*, *Pseudomonas*, *Bacillus*, and even new unnamed isolates. We used microorganisms to make a panoply of products from interleukins for treating cancer to phenylalanine for manufacturing the sweetener aspartame to enzymes for cleaning drains. Although I was with the company only a short time (1981 to 1984), the thrilling pace and range of experiences made it feel the equivalent of much longer.

When the president of Penn State University sought me out in 1984 to found a new biotechnology institute, I accepted the offer as an opportunity to couple my academic experience with this new enthusiasm for using microorganisms in biotechnology. My role as institute creator, fundraiser, recruiter, building designer, and educator gave me yet another view of biotechnology. Through this larger administrative lens, I could see that there were certain problems shared by many biotechnology companies—common bottlenecks—that kept ideas from ever becoming products. I viewed a biotechnology institute as a place where researchers could work to solve these common problems and open the doors for even greater successes. Simultaneously, through my role as president of the American Society for Microbiology (1986), I saw the need to become a spokesperson and explain this rapidly growing, highly publicized world of biotechnology to legislators, industry leaders, and the public.

In 1991, with the launching of the institute complete, I faced the choice of continuing in academic administration, perhaps as a university president somewhere, or returning to my roots in research and teaching.

I chose to leave the administrative world to pursue my own research vision of putting microorganisms to work. Whereas most biotechnology companies had focused on making high-value human pharmaceutical products, I saw an untapped treasure hiding in the varied physiology of diverse microorganisms. Previously, many microorganisms making useful metabolites and enzymes were thrown away because they grew too slowly or produced too little to make commercialization feasible. Now, we could hurdle those barriers by cloning the desired genes into organisms that can be grown rapidly in huge fermenters.

The usefulness of this approach had been demonstrated by the cloning of genes from thermophiles into new hosts for the production of heat-stable enzymes, such as the Taq polymerase used in polymerase chain reaction. Researchers began looking for more examples of useful thermophilic enzymes. At the other end of the temperature scale, however, psychrophilic (cold-loving) microorganisms had been largely neglected. One criterion I had for selecting my new research interest was that it should help champion an understudied, but important, facet of microbiology. The isolation of psychrophiles and the study of their cold-active enzymes with high activity at low temperatures filled that need.

The many cold, chemically unique environments on Earth promised a rich diversity of psychrophiles waiting to be revealed. The study of cold-active enzymes could lend insight into the basic features that set an enzyme's thermostat for activity. And, in keeping with my heritage in biotechnology, cold-active enzymes had potential applications. Cold-active proteases and pectinases could improve meat and beverage processing at low temperatures, and β-galactosidases could hydrolyze lactose in refrigerated milk, making it drinkable by the majority of the world's population who are lactose intolerant. Cold-active enzymes could make more powerful detergents for low-temperature cleaning, they could catalyze reactions with volatile chemicals, or they could degrade hazardous wastes in cold climates.

As part of this new endeavor, my research group has isolated several hundred psychrophiles from samples gathered in cold-climate soils, springs, lakes, oceans, and, yes, the Antarctic. Some isolates belong to previously undiscovered bacterial genera and species based on their physiological properties and 16S ribosomal RNA gene sequences. We screen our isolates for cold-active enzymes such as glycosidases, phosphatases, and proteases. In one case, we cloned three different genes, each encoding an unusual β-galactosidase, from one *Arthrobacter* isolate. We have now cloned, sequenced, and phylogenetically compared about twenty different glycosidase genes from a variety of isolates. The discovery of these genes adds to the fundamental knowledge of glycosidase functions and the evolution of these activities and yields potentially useful cold-active enzymes. In addition, we are striving to learn how protein sequences and structures affect activity at different temperatures by comparing cold-active enzyme structures with their higher-temperature counterparts from mesophiles and thermophiles.

The study of psychrophilic organisms is just one example of an understudied but important area just waiting for future microbiologists to probe. So little do we yet know about diverse forms of life around us that less than 1% of observed microbes have been isolated. Microbiology offers great opportunities for future students eager to be pioneers exploring the microbial world. Discoveries through basic research of unique lifestyles continually amaze us, and new biotechnology products await adventurers with the skill to develop them. As I hope my experience illustrates, microbiology can open the curtain on a varied and entertaining career. Microbiology can escort students along paths through universities, industries, and clinics as researchers, managers, and administrators. My path has taken me from childhood mysteries of science on the farm to basic findings of unknown pathways, into the world of biotechnology and administration, and on to discovering new organisms and useful enzymes.

JEAN BRENCHLEY was born in Towanda, Pennsylvania, in 1944. She received her bachelor of science degree from Mansfield State College in 1962, her master's degree from the University of California, San Diego, in 1967, and her doctoral degree from the University of California, Davis, in microbiology in 1970. She did postdoctoral studies at MIT, then joined the faculty at Penn State in 1971, and was later recruited to Purdue University. In 1981, she left Purdue to become research director in a small biotechnology company. Then in 1984 she became Head of the Department of Molecular and Cell Biology and Director of the Biotechnology Institute at Penn State University. In 1990 she became Professor of Microbiology and Biotechnology at Penn State University. She was President of the American Society for Microbiology in 1986. Recognition of her professional service and research accomplishments in microbial genetics, biotechnology, and the discovery of novel cold-active enzymes have included the Waksman award, Alice Evans Award, membership on the National Biotechnology Policy Board, and service as an editor to various journals.

The following papers are representative of her publications:

Coombs, J.M. and J.E. Brenchley. 1999. Biochemical and phylogenetic analyses of a cold-active β-galactosidase from the lactic acid bacterium Carnobacterium piscicola. BA. *Appl. Environ. Microbiol.* **65:**5443–5450.

Loveland-Curtze, J., P. P. Sheridan, K. R. Gutshall, and J. E. Brenchley. 1999. Biochemical and phylogenetic analyses of psychrophilic isolates belonging to the *Arthrobacter* subgroup and description of *Arthrobacter psychrolactophilus*, sp. nov. *Arch. Microbiol.* **171:**355–363.

Gutshall, K., K. Wang, and J. Brenchley. 1997. A novel *Arthrobacter* β-galactosidase with homology to eucaryotic β-galactosidases. *J. Bacteriol.* **179:**3064–3067.

Brenchley, J. E. 1996. Psychrophilic microorganisms and their cold-active enzymes. *J. Indust. Microbiol. Biotechnol.* **17:**432–437.

Gutshall, K., D. Trimbur, J. Kasmir, and J. Brenchley. 1995. Analysis of a novel gene and β-galactosidase isozyme from a psychrotrophic *Arthrobacter* isolate. *J. Bacteriol.* **177:**1981–1988.

Trimbur, D., K. Gutshall, P. Prema, and J. Brenchley. 1994. Characterization of a psychrotrophic *Arthrobacter* gene and its cold-active β-galactosidase. *Appl. Environ. Microbiol.* **60:**4544–4552.

Loveland, J., K. Gutshall, J. Kasmir, P. Prema, and J. Brenchley. 1994. Characterization of psychrotrophic microorganisms producing β-galactosidase activities. *Appl. Environ. Microbiol.* **60:**12–18.

29

JENNIE HUNTER-CEVERA

Following Diverse Vistas

What are you doing out there for so long? Why are you staring at that piece of grass? These were the questions that my mother would yell out the screen door on those hot, muggy July days in Monessen, Pennsylvania. I used to lie out in the sun on a beach towel trying to get the perfect summer tan as a young teenager and to imagine what was going on inside a blade of grass. For a while my mother thought I needed counseling because I was more fascinated with the colors and textures of grass and soil than in the top ten records. This concern was not new. In the fourth grade I had organized a science club. I felt it was important that everyone know about science. We were growing up in the nuclear and space age with so many new technologies coming on line that would make our lives simpler and more efficient and allow more time for fun. It seemed that there was so much to learn and not enough time. I always wanted to find out why things—especially living things—worked the way that they did.

Growing up in an Italian-Scottish household was not an easy task but certainly a memorable experience. Between haggis and lasagna, single-malt scotches and Chianti, scones and biscotti, it is no small wonder that most of my life and career have always provided me with two choices, two paths to follow, often resulting in a hybrid life. It also provided me with a sense of values and morals that has stayed with me throughout my professional and private life. My parents taught me that knowledge was something that no one could take away from you. They could take your car, your wallet, your job, but not your knowledge.

Besides always having my head stuck in a book, I loved to draw and paint with watercolors. In my senior year I had to decide between a career in art or biology. I was temperamental about my artwork. I would often stay up for days until a painting was completed. Therefore, I decided that a career in medicine might better suit my personality. Besides, my parents expected me to follow in my grandfather's footsteps and become a doctor.

Penn State and West Virginia University (WVU) had accepted my application to attend college and major in premedicine. Deciding which college to attend was not easy because many of my friends were going to Penn State. When I visited the WVU campus in Morgantown and saw the curriculum, however, I knew I wanted to be a Mountaineer. It was the right choice because all the science courses had lab components. You also had to take several nonscience courses to "round you out." Many of these nonscience courses gave me a new perspective on life and choices.

Majoring in premed required that I take a microbiology course in my senior year. The first time I saw a bacterium under the microscope, I became addicted to studying these invisible keepers of the earth. Hal Wilson was the course instructor, and he knew early on that I was falling in love with microbiology. He offered me a senior internship to work with a Ph.D. graduate student named Fred Thompson. Fred was studying the effects of acid mine drainage on the normal microbial biota found in the Monongahela River. I loved working in the laboratory as a research technician. Even though I had been interviewing at medical schools, I decided to go to graduate school at WVU to obtain a master's degree in microbiology. It was 1970, a time when people were becoming more aware of pollution and contamination in the environment. I decided to forgo medicine and become more involved in trying to clean up the environment.

Wayne Millar was a new professor at WVU in the Department of Plant Pathology and Bacteriology. He had a research grant to explore the use of biological agents to clean up acid mine drainage. I became his graduate student and tried to isolate strains of *Bdellovibrio bacteri-*

ovorus that might be active against members of the genus *Thiobacillus*. *Bdellovibrio* was a new genus discovered by Drs. Stolp and Starr. It was the only known parasitic bacterium. Its hosts were mainly gram negative, such as *Escherichia coli*. I sampled many a sewage plant in West Virginia and surrounding soils with low pH in hopes of finding the "magic bullet" that would attack *Thiobacillus ferrooxidans*. I was successful in isolating a strain of *B. bacteriovorus* that attacked *T. novellus* but not *T. ferrooxidans*, because at low pH the flagellum of *B. bacteriovorus* fell off and disabled the parasite from moving at a fast speed. Investigators at the time thought that the combination of hitting the host at a fast speed and producing enzymes that bored into the cell wall were the mechanisms that enabled *B. bacteriovorus* to attack host cells. Unfortunately, I never presented this study to the scientific community and Dr. Millar moved on to a research position at Eli Lilly and Company. I realized later in my career that I should have published this work.

During these two years at WVU, I was exposed to many different areas of microbiology (soil, food, sewage, agriculture, pathology, and ecology) and decided to pursue a career in applied microbiology. Other graduate students were either staying at WVU or moving to other universities to get Ph.Ds. Drs. Lilly and Barnett, my instructors for mycology, tried to convince me that I should continue my education. I was unsure that I needed a Ph.D. to work in industry, so I decided instead to hunt for a job.

It took five months of sending out letters, mainly to chemical and food companies, and interviewing before I landed what I considered a dream job. Meanwhile, I had been working in the local flower shop earning a small salary and wondering if I had made a mistake by not staying in school. My mother knew I was unhappy and encouraged me to write pharmaceutical companies because she believed that the health care industry would offer a good career for me. So I did. The irony of landing this dream job was that I wrote E. R. Squibb and Sons inquiring about employment opportunities not because I wanted to work for a pharmaceutical company but because my mother took their vitamins daily and thought that Squibb was a great company. I actually obtained the address from a jar of vitamins.

E. R. Squibb and Sons offered a research assistant position to me in May 1973. My responsibilities were to isolate rare actinomycetes (nonstreptomycetes) from soil samples that were sent to Squibb from all over the world. These actinomycetes would be grown in liquid broth, extracted, and assayed for novel antibiotic activity against pathogenic microorganisms. Squibb's facilities were brand new and I had a lab bench all to myself! There were so many exciting new areas of research to learn about, and a great team of scientists, including my supervisor William Trejo, surrounded me. He was the world's expert on blue-spored streptomycetes. He was also an accomplished classical guitarist. (I have found that most scientists are also talented artists, musicians, or writers.) I thought that I had died and gone to research heaven. This was the beginning of my professional career as an industrial microbiologist.

Six months later, I realized that I needed a "union card" if anyone was going to take my ideas seriously. I began to apply to Ph.D. programs. Meanwhile, I wrote a report on a set of experiments that I did showing that our cell wall screen had some flaws. I distributed the report to the senior investigators as well as my peers. I had upset the apple cart because junior staff members never sent out reports. I was immediately called into the director's office. I thought for sure that I was going to be fired on the spot. Rather than be fired, I would resign. Much to my surprise and delight, the director thought my experiments made a lot of sense and offered me a golden opportunity—to attend graduate school at Rutgers University and have Squibb foot the entire bill!

My Ph.D. thesis documented the presence and activity of actinomycetes in a salt marsh. Squibb would have access to screening the salt marsh isolates and I would screen them for enzymatic activity. This way there was no conflict of interest. Douglas Eveleigh was my thesis advisor and Hubert Lechevalier and his wife, Mary Lechevalier, became my mentors on isolating and identifying actinomycetes. I was experiencing the best and worst of both academia and industry at the same time. Even though I was a full-time graduate student, I still had research responsibilities to carry out for the soil screens at Squibb. I was given several assistants and was put in charge of soil isolation work. I traveled to England and visited the laboratories of Tom Cross in Bradford and Michael Goodfellow in Newcastle upon Tyne. I ate, lived, and slept with actinomycetes for four years.

I was convinced that novel actinomycetes and novel secondary metabolic activity could be found if one included the ecophysiologic parameters of the sample material into the isolation media and screening methods. This hypothesis was tested when I isolated *Chromobacterium vioacleum* from the Pine Barrens by using a low pH medium. When grown in slightly lower pH medium than usual, it produced a novel cell wall inhibitor compound called a *monobactam*. Four years later, after much hard work, I had a Ph.D. in microbiology and E. R. Squibb and Sons had a new antibiotic, azthreonam I realized that even though I had a Ph.D. now, it didn't matter because my colleagues thought of me as what I started at Squibb rather than what I'd accomplished. I left Squibb just as the papers were beginning to surface in the literature about the monobactams. My name appeared on only one of these papers and others took credit for my isolation work. A hard lesson was learned, but it enabled me to think "outside of the box" and move on after being with Squibb for almost eight years.

Cetus Corporation was a young biotechnology company in Berkeley, California. I wrote the chief executive officer, Ronald Cape, looking for a job. There was a group within the company known as New Ventures. Charles Cooney at MIT knew of their interest in discovering novel enzymes and thought that I would be a good fit. Even though I had several job offers from European pharmaceutical companies and Cetus was working mainly out of a warehouse, I knew on my initial visit to the 5th Street site in Berkeley that I wanted to be part of the multidisciplined, diverse Cetus research team. This was a hard decision because I just had become engaged to my future husband, Ray Cevera, who also had five children still living in New Jersey. We decided to go west and make our home there. I began at Cetus on Halloween day in 1980. Ray and I were married in May 1981. Eventually the children came to live with us in California. Thus, I started a new job, got married, and became an instant mother—all within a year.

Cetus turned out to be the greatest adventure of my career in microbiology. Not only did we discover novel haloperoxidases using an ecological approach for isolation and screening, but also we worked on several applied research projects at the same time, such as converting ethanol into ethylene; making ascorbic acid via one-step fermentation, enzymatic breath mints, recombinant beer, and improved naturally sweetened cereals; converting coal dust to single-cell protein; using isosaccharinic acid from kraft waste liquors to make chemical commodities; as well as scaling up procedures for interleukin-2 and β-interferon. I established the Cetus culture collection. I also had two children of my own, Kristen and Jonathan. Juggling work and motherhood became a way of life.

My love affair with Cetus lasted for 10 years until I was laid off because interleukin-2 was not approved by the U.S. Food and Drug Administration (FDA). I was devastated and depressed because I thought Cetus would be my research home forever. But nothing lasts forever, and I had to pick myself up off the ground and start over looking for a new job.

Meanwhile, Angela Belt, the curator of the Cetus culture collection, thought we should set up a small consulting company to assist young biotechnology companies in getting ready for FDA inspections, setting up culture collections, and identifying actinomycetes and fungi. The Biotic Network was formed and soon led to the founding of Blue Sky Research Service, which specialized in designing fermentation conditions for unusual or novel microorganisms and developing new extraction methods. For one year, I traveled 2.5 hours to work in Sonora, California, for days at a time and then went back to El Cerrito to consult for companies in the San Francisco Bay area and in Europe. The Biotic Network was a "mom and pop" operation, but it enabled me to see what else was going on in biotechnology after being with one company for so long. During this period, I also worked part-time at Biosource Technologies with Barry Holtz on finding ways to decrease the compost prep time for mushroom farms and increasing the flavor of mushrooms early in their growth stages. Thermophilic actinomycetes proved to be the key ingredient for a compost inoculum.

Although I loved the independence of being on my own, the up and down income was stressful. I took a position as Research Fellow at Geobiotics in September 1991. It was a small biotech company in Hayward, California. The mining industry had a problem with tailings containing precious ores and wanted a cost-effective way to remove them. William Kohr, President of Geobiotics, wanted to explore the use of microorganisms and enzymes to extract the precious ores, as well as deal with cyanide waste. In one year, we isolated strains of *Xanthomonas* from cyanide pits that were able to use the cyanide as a carbon and/or nitrogen source and bind gold. Unfortunately, the mining industry was not really ready for biotechnology in 1992 and preferred incineration to the power of microbes. I was worried about the stability of the company and decided that perhaps it was time to leave California because Ray was facing a possible layoff.

I accepted a position as Associate Director at Universal Flavors in Indianapolis, Indiana. I was to build a biotech group that would discover enzymes and metabolites that could be used in the food and flavor business and optimize scale-up processes for the manufacturing of natural flavors. It was my introduction to the flavor business and I loved the novelty of it all. To work with flavor chemists is a surreal adventure. They are incredibly bright and have uncanny taste buds that can decipher ingredients and their ratio better than a high-pressure liquid chromatography machine. I learned so much in nine months, and my team worked on natural butter flavor for kosher popcorn, a new natural grape flavor, natural production of sulfur compounds, natural beef flavors, and peach lactones.

There were internal management problems brewing at Universal's headquarters in Madison, Wisconsin, and I decided that there was a lack of commitment to biotechnology for flavor production. I was right; the group was disbanded a year after I left. I went back to California where my husband was still employed and my children were attending school. However, the day I drove into my driveway was the day my husband received his layoff notice. Fate took a twisted turn and this time it was double jeopardy.

In my search for a new position, I called a former Cetus employee, Sharon Shoemaker, who was now the Executive Director of the California Institute of Food and Agricultural Research at the University of California at Davis. Sharon needed some help dealing with the Phaff yeast collection and in setting up a new initiative on neutraceuticals. I agreed to work part-time while I still consulted for the Biotic Network and did fermentation work for Blue Sky Research. Other than being a student, this was my first introduction to working in academia. Working with Dr. Phaff, who was by now in his mid-80s, inspired me to always stay connected to the laboratory. The best research project manager is one who understands and knows what is going on in the laboratory. We obtained funding from the National Science Foundation that saved Dr. Phaff's valuable yeast collection.

Consulting eventually led me to the Lawrence Berkeley National Laboratory (LBNL). I had never considered working for a national laboratory mainly because I knew little about them. LBNL wanted to set up a Center for Environmental Biotechnology (CEB) that would link the many different people working in this area at LBNL and the University of California at Berkeley. I decided to interview for the position and was hired on November 4, 1994. Here was an opportunity for me to shape the CEB and develop integrated research programs that would link physics, chemistry, and biology together to find new solutions to old problems. It was a new concept for both LBNL and Berkeley to form teams of multidisciplined researchers and to study microorganisms as they functioned under conditions found in nature instead of the laboratory.

In the fall of 1996, CEB became involved with nonproliferation, including biological weapons, and I wanted to see if a DNA assay could be developed to detect *Bacillus anthracis*. Nancy McKinney found that the small acid-soluble protein (SaspB) was unique for *B. anthracis* and consistently identified it from its near neighbors, *B. cereus* and *B. thuringiensis*. We had found the first chromosomal marker for *B. anthracis*. Today, CEB is thriving and has three research focus areas: environmental diagnostics for water quality, biowarfare diagnostics, and environmental monitoring; biogeochemical transformations for bioremediation and synchrotron assessment and validation; and environmental risk assessment based on bioavailability through ingestion, bioabsorption, and inhalation.

I am an advocate for biodiversity and culture collections, as well as a big believer that nature is by far the best genetic engineer. Natural products will always play an important role in medicine, food, and the environment. I think that the twenty-first century will bring more integration of physiology and molecular biology and use the physicists' tools more to elucidate and validate the structure and function of microorganisms in our world.

In the summer of 1999, I accepted the challenging position of President of the University of Maryland's Biotechnology Institute. I am going home, back to the East Coast, where my career began. In some ways, it seems that all of my past working experiences in industry, academia, and government have paved the way for this new position. We are all products of our environment and life experiences; I am a hybrid between very different worlds starting with the world I was raised in by my parents to my talents developed by my professional experiences. I look forward to being part of future discoveries, new products and processes, and a team that will continue to think outside of the box. I am still curious about the invisible creatures that seem to rule the earth and even our lives in many ways. I feel fortunate to have worked in so many different areas of microbiology. I have never been bored in my career—just overworked!

JENNIE HUNTER-CEVERA was raised in Monessen, Pennsylvania. In 1970 she graduated from West Virginia University with a bachelor's degree in biology. She obtained her master's degree from West Virginia University in microbiology in 1972. In 1973 she was employed as Research Assistant at E. R. Squibb and Sons in New Jersey. She received her doctoral degree in microbiology from Rutgers University in 1978, having worked at Squibb throughout her graduate studies. As part of her graduate work she isolated a strain of *Chromobacterium violaceum* from the Pine Barrens in New Jersey that produced a novel antimicrobial structure—monobactams. She took a position with Cetus Corporation in 1980 and the following year discovered novel haloperoxidases from Death Valley. She cofounded The Biotic Network in 1990. In 1991, while employed by Geokinetics, she isolated cyanide-degrading bacteria. In 1992 she worked for Universal Flavors, forming a Biotechnology Department. In 1994 she moved to Lawrence Berkeley National Laboratory and developed the Center for Environmental Biotechnology. In 1999 she accepted the position of President of the University of Maryland's Biotechnology Institute.

The following books and papers are representative of her publications:

Holman, H.-Y., M. Zhang, R. Goth-Goldstein, M. C. Martin, M. Russell, W. R. McKinney, M. Ferrari, and J. C. Hunter-Cevera. 1999. Detecting exposure to environmental organic toxins in individual cells: Towards development of a micro-fabricated device. *SPIE* **3606:**55–62.

Holman, H.-Y., D. L. Perry, and J. C. Hunter-Cevera. 1998. Surface-enhanced infrared absorption-reflectance (SEIRA) microspectroscopy for bacteria localization in geologic material surfaces. *J. Microbiol. Meth.* **34:**59–71.

Hunter-Cevera, J. C. 1998. The value of microbial diversity. *Curr. Opin. Microbiol.* **1:**278–285.

Hunter-Cevera, J. C., and A. Belt. 1996. *Maintaining Cultures for Biotechnology and Industry*. Academic Press, San Diego.

Hunter, J. C., A. Belt, and A. P. Halluin. 1986. Guidelines for establishing a culture collection within a biotechnology company. *Trends Biotechnol.* **4:**5–11.

Hunter-Cevera, J. C., and L. Sotos. 1986. Screening for a "new" enzyme in nature: Haloperoxidase production by Death Valley dematiaceous hyphomycetes. *Microb. Ecol.* **12:**121–127.

Hunter, J. C., M. Fonda, L. Sotos, R. Toso, and A. Belt. 1983. Ecological approaches to isolation. *Dev. Indust. Microbiol.* **25:**247–266.

Wells, J. S., J. C. Hunter, G. Astle, J. Sherwood, C. M. Ricca, W. H. Trejo, D. P. Bonner, and R. B. Sykes. 1982. Distribution of β-lactam and β-lactone producing bacteria in nature. *J. Antibiot.* **35:**814–821.

Hunter, J. C., D. E. Eveleigh, and G. Casella. 1979. Actinomycetes of a salt marsh. *Zbl. Bakt. Suppl.* **11:**195–200.

A. M. CHAKRABARTY

Moving Ahead with Pseudomonad Genes

My career has revolved around the pseudomonads, moving my interests from one *Pseudomonas* species to another, speedily moving from metabolism to genetics, and in the course of my career in industry and academia, trying to pursue earnestly the science underpinning biotechnology during its explosive development of the last few years.

My romance with the pseudomonads began during my Ph.D. studies at the University of Calcutta, India, where I studied the nature of pigments synthesized by the *P. fluorescens* group. I then moved to the University of Illinois at Urbana-Champaign to continue studying pseudomonads as a postdoc in the laboratory of I. C. Gunsalus. I had no real problem in settling down because Urbana was (and still is) a truly international campus where foreign students abound. I still recall with pleasure that when my wife and I arrived in Urbana, we had about $32 between us. We arrived in Urbana at about 8 P.M. After a few minutes of figuring out how to get to the Illini Union, which Dr. Gunsalus's secretary asked us to do when we called the office from Chicago, we were approached by a tall, handsome gentleman who introduced himself as Mr. John Price from the local YMCA. Mr. Price had a car, and his intention was to provide rides to newly arrived foreign students who had either no transportation or no place to go. He not only dropped us at the Illini Union, but on many occasions, particularly during Christmas holidays when nobody was around, he took us to Chicago to visit the museums, gave us warm clothing for the winter, and helped us find hosts on Thanksgiving holidays. The selfless and dedicated service provided to newcomers in the University town by Mr. Price and his colleagues touched us then and still remains in my memory as a shining beacon of what a true Christian attitude and welcome are like. I certainly have not forgotten his name, even after 33 years.

At the University of Illinois I quickly moved to my work in Dr. Gunsalus' laboratory on the isolation and characterization of transducing phages for *P. putida*. Dr. Gunsalus—known as Gunny—was extremely interested in knowing how and why *P. putida* exhibited a broad degradative capability. During my postdoctoral years, we came up with some very exciting observations in rapid succession, which showed that in *P. putida*, the genes for the degradation of rather exotic organic compounds—such as camphor, octane, and naphthalene—occur on plasmids. Interestingly, we found that the genes specifying each of the degradation pathways occurred on individual plasmids. Before our work, such pathways were known to be specified by chromosomal genes. We had made a significant advance in our basic understanding about the genes involved in the biodegradation of exotic compounds, one that also would have significant practical implications.

In 1971, I moved to the Corporate Research and Development Center of General Electric (GE) at Schenectady in upstate New York. There I learned to think about practical applications of day-to-day research. In the early 1970s, environmental consciousness was growing and environmental laws were becoming stringent. GE was interested in many aspects of microorganisms and environmental pollutants. Feedlot owners were under pressure to develop alternative means of disposing cattle manure, and a major constituent of manure is lignin and its cellulosic complex derived from undigested plant materials. I began working on lignocellulose degradation with Dex Bellamy, who had the bright idea of digesting manure lignocellulosic material by thermophilic actinomycetes.

But I could not completely forget my pseudomonads. During my off-hours and weekends, I studied the abilities of pseudomonads to degrade hydrocarbons through plasmid-encoded pathways. I noticed that pseudomonads were pretty versatile in degrading a variety of petroleum hydrocarbons. At this time crude oil was cheap, and there was talk of converting oil to protein that could be used as cattle feed. I became interested in developing a process of converting oil to bacterial proteins by using pseudomonads.

But I soon hit a snag. Crude oil is a mixture of many hydrocarbons, but individual pseudomonad strains had only a limited number of degradative plasmids that encoded hydrocarbon degradation. Thus in order to accomplish significant crude oil consumption, I needed to use a mixture of pseudomonads, each capable of degrading different hydrocarbons. Unfortunately pseudomonads are like people; they cannot tolerate one another. Thus a mixed culture growth always resulted in the elimination of some strains and the dominance of others. It occurred to me that because the pathways were encoded by plasmids, it might be possible to transfer different hydrocarbon degradative plasmids into a single strain.

Having worked on plasmids specifying degradation of terpenes, alkanes, and so forth during my postdoctoral days at Urbana, I started designing a multiplasmid pseudomonad harboring a number of hydrocarbon-degradative plasmids when I came to work for GE at Schenectady. Some quick testing for growth on crude oil demonstrated that a multiplasmid bacterium grew faster and better on crude oil than its single-plasmid parents. I decided to talk about this finding in a meeting at Tel Aviv, Israel, where I was invited. To make a presentation at an international meeting, I needed approval from my management, including the patent lawyers. I submitted an abstract of my talk to my manager, hoping to get a routine approval. A few weeks before I was to leave for Israel, I was informed that the vice president, Arthur Bueche, thought that the ideas were intriguing and that I should get a patent before talking publicly about my findings. This took me completely by surprise. I did not know much about patents, except that if GE chose to file for a patent, I would get $250. Normally I would be happy with the management decision, except I was committed to talk at the Tel Aviv meeting, and the management decision put a stop to my talk. When I complained to my manager, he asked me to attend the meeting in Israel but not to present my talk. Soon afterward a patent was filed on my behalf.

The patent covered both the process of making a multiplasmid organism as well as the multiplasmid pseudomonad itself. The patent office (PTO) agreed on the process patent but denied the claim on the bacterium, arguing that living bacteria, since they are products of nature, are not patentable. GE appealed the decision to the court of Customs and Patent Appeals (CCPA), pointing out that a genetically engineered bacterium is not a product of nature but a genetically improved, well-designed construct that is novel and useful. The CCPA agreed and overruled the PTO decision. The PTO then appealed to the U.S. Supreme Court, which sent it back to the CCPA, asking them to reconsider. The CCPA ruled in GE's favor for a second time. The PTO again appealed to the Supreme Court, leading to a dramatic court battle over the patentability of life forms. By then I had left GE to accept a position at the University of Illinois in Chicago. However, I frequently consulted with Ed MacKey, the GE lawyer who argued the case on our behalf in the Supreme Court. Sydney Diamond, as a Commissioner of Patents, brought the case for the government and the solicitor general argued their side in the court. I attended the court session, which was very dramatic at times, because of the sharp questioning to both sides from some of the justices. Eventually the court ruled 5-4 in GE's favor, and patenting of life forms became a reality. This was truly a landmark decision that opened the door for major investments in biotechnology and increased the rivalry among the emerging biotechnology companies.

While the Supreme Court ruling settled the case over the patentability of bacteria, many people remained uneasy about the implications of that decision. Many felt that patenting altered life forms, though starting out with a bacterium, could end in the patenting of higher forms of life, perhaps even humans. That has in fact happened, at least to some extent. Plants, animals, human cell lines, and human disease genes with single-nucleotide polymorphisms have all been patented. Significant questions have been raised regarding the impact of the initial Supreme Court decision on subsequent events. Indeed, the impact of human genetic intervention, including the intellectual property rights issues and the propriety of expressing human genes in animals, the potential production of human organs in animals, or even attempts to construct potential human-animal hybrids, is becoming an issue of great concern and significance. I often meet with members of the judiciary, advising judges on the impact of advances in human genetics. Judges need to be prepared for the cases they soon will face.

Returning to science, in the mid-1970s, my interest moved from petroleum hydrocarbons to polychlorinated biphenyls (PCBs). The

State of New York accused GE of contaminating the Hudson River with PCBs. Even though GE had a permit to use PCBs as dielectric fluids in the electrical transformers, the PCBs somehow ended up in the bottom of the Hudson River. GE agreed to contribute to the cleanup of PCBs from the river sediment, as part of a legal settlement. I was assigned to work on the metabolism of PCBs by microorganisms in Hudson River sediments. The release of many chlorinated aromatics to the environment in massive amounts as herbicides or pesticides raised an important question: Do natural microorganisms, having been exposed to such synthetic compounds, evolve new genes whose products (the enzymes) can allow complete or nearly complete degradation of these compounds? Indeed, while working at GE, we demonstrated the presence of a plasmid encoding a *p*-chlorobiphenyl degradative pathway in a Hudson River sediment isolate. It was at this time, however, that I got an offer from the University of Illinois at Chicago to join its faculty. In 1979, when I moved to Chicago, I maintained an active interest in the microbial evolution of new biological functions, particularly the degradation of synthetic toxic chemicals.

At the time I joined the University of Illinois at Chicago, biotechnology was poised to move forward with explosive speed—and so it has. Herb Boyer and his group had already cloned and expressed the somatostatin gene in *Escherichia coli*, and his group, as well as that of Bill Rutter, were busy cloning and expressing the insulin gene in *E. coli* to replace animal-derived insulin. My previous experience with the construction of multiplasmid *Pseudomonas* for oil consumption and the occurrence of PCB-dissimilating organisms isolated from Hudson River sediment convinced me that the evolution of new degradative genes for the dissimilation of synthetic chlorinated compounds was in full swing in nature. If the process of natural evolution of these genes could be understood, it could be enhanced by genetic manipulation and exploited. With the emerging genetic engineering techniques, we could begin to construct engineered microorganisms that could consume toxic chemicals and chelate or biotransform heavy metals to reduce their toxicity. With these goals in mind, my students, postdoctoral colleagues, and I embarked on studying various facets of microbial degradation of synthetic, chlorinated compounds, including the chemostatic selection of microorganisms capable of degrading a normally recalcitrant compound such as 2,4,5-trichlorophenoxy acetic acid. The goal has been to understand how microorganisms, particularly members of the *Pseudomonas* species, recruit new catabolic genes and organize them so that the genes are well expressed and well regulated.

Although bioremediation has been a popular subject for discussion, its practical applications have been few and far between. Most contaminated sites have multiple contaminants, often a mixture of toxic chemicals and heavy metals. Microorganisms, even genetically engineered ones, have only limited biodegradative capability, so that only a few contaminants can be decomposed at a time. The presence of other contaminants tends to inhibit the growth of the bioremediative organisms. Also, there are natural substrates in soil, such as lignocellulosic materials, that are often preferred by the organisms rather than the toxic chemicals, whose effective concentrations, such as that of 2,3,7,8-tetrachloro-dibenzo-*p*-dioxin, might be extremely low to be of meaningful nutritional value to the organisms. Although these are technical problems that can be addressed, a major hurdle in the development of microbial bioremediation has been the regulatory considerations. Genetically engineered organisms that are grown in fermenters for the production of biomedically important animal or human proteins, or transgenic plants producing desirable products, can be quickly destroyed or burned if something goes wrong. Bioremediative microorganisms, once released, cannot be reached and destroyed, because they are blown all over by the wind. There is no recall for environmentally released organisms, genetically engineered or not. Thus, a major question in the release of genetically engineered microorganisms, or even indigenous microorganisms, is their pathogenic potential. Could the released microorganisms, which should have a high growth advantage in a contaminated area, acquire virulence factors from the neighboring organisms that could make them pathogenic? Alternatively, could the degradative genes be transferred from the engineered organisms to a pathogenic neighbor, who could then grow well because of the acquisition of the degradative genes and the availability of the substrates in the contaminated site? Is the environment of the contaminated site conducive to the promotion or expression of microbial pathogenic potential?

Being in a medical center, I was acutely aware of the havoc pathogenic microorganisms can create. I teach the molecular basis of infectious diseases to the Ph.D. and M.D. students and have a keen interest in understanding how the environment modulates microbial pathogenicity. Because most of the bioremediative organisms are *Pseudomonas* or closely related organisms, and since some pseudomonads, like *P. aeruginosa* and *P. syringae*, are human and plant pathogens, I became involved in studies of the molecular basis of *P. aeruginosa* infection in the lungs of cystic fibrosis patients and later on respiratory infections by slow-growing mycobacteria, such as M. *bovis* or M. *tuberculosis*. The early studies of *Pseudomonas* degradative plasmids and their applications and roles in bioremediation propelled me to study their mode of pathogenicity and the nature of host-pathogen communication that dictate whether the host will be able to restrict the pathogen. Such mechanisms, observed first in *Pseudomonas*, appear to be universal and may contribute to our understanding of how other pathogens take over the host cell machinery. It is hoped that our studies will contribute to our ability to limit *Pseudomonas* infections in children with cystic fibrosis.

A final note about being from India and working within the community of microbiologists in the United States. Friends and colleagues often ask me about discrimination. Many complain about discriminatory hiring practices and hostility in the work and marketplaces toward foreigners and individuals of minority racial and ethnic groups. I remember vividly a dinner conversation with one of my faculty friends in which I was asked pointedly if I ever felt discriminated against. I thought very seriously and answered with a firm no. Within my scientific career, whenever I was denied a grant or my paper was rejected by a journal—and that has happened to me—I would ask myself what had gone wrong and how I could improve my performance. I never ascribed rejection to discrimination; if it occurred I was blind to it. Then I asked her—a charming, highly successful, young Caucasian woman—if she ever felt she had been discriminated against. She answered yes—that as a woman, she felt that she often was the victim of gender bias. Her answer was as surprising as it was revealing. Discrimination takes various forms and shapes and transcends gender, age, color, and national boundaries. I have been fortunate to be able to pursue my career as a microbiologist unhindered by such limitations. My lifelong study of the pseudomonads has been an exciting and rewarding experience. I fancy myself spending the rest of my academic career pursuing the intriguing lifestyle and the fascinating activities of the pseudomonads—and hopefully moving science and technology ahead.

A. M. CHAKRABARTY was born in 1938 in Sainthia, India. He received his bachelor of science degree in 1958 from St. Xavier's College in India. He received a master's degree from Calcutta University in 1960 and a doctoral degree in biochemistry from Calcutta University in 1965. He has an honorary D.Sc. degree from the University of Burdwan. He was a research scientist at General Electric before becoming Professor of Microbiology in 1979 at the University of Illinois College of Medicine, where he currently is Distinguished Professor. He has received the Scientist of the Year Award and Pasteur Award and the Procter and Gamble Award. His research focuses on the evolution and application of hydrocarbon degradative plasmids in *Pseudomonas*, molecular cloning and genetic engineering for biodegradation of environmental pollutants, and *Pseudomonas* infection in cystic fibrosis.

The following books and papers are representative of his many publications:

Zaborina, O., X. Li, G. Cheng, V. Kapatral, and A. M. Chakrabarty. 1999. Secretion of ATP-utilizing enzymes, nucleoside diphosphate kinase and ATPase, by *Mycobacterium bovis* BCG: Sequestration of ATP from macrophage P2Z receptors? *Mol. Microbiol.* **31:**1333–1343.

McFall, S. M., S. A. Chugani, and A. M. Chakrabarty. 1998. Transcriptional activation of the catechol and chlorocatechol operons: Variations on a theme. *Gene* **223:**257–267.

Chakrabarty, A. M. 1998. Diamond v. Chakrabarty: A historical perspective. *In* D. S. Chisum, C. A. Nard, H. F. Schwartz, P. Newman, and F. S. Kieff, (ed.), *Principles of Patent Law*. Foundation Press, New York, N.Y.

Chakrabarty, A. M. 1996. Microbial degradation of toxic chemicals: Evolutionary insights and practical considerations. *ASM News* **62:**130–137.

Chakrabarty, A. M. 1982. Plasmids and nutritional diversity. *In* L. N. Ornston and S. G. Sligar (ed.), *Experiences in Biochemical Perception*. Academic Press, New York, N.Y.

Chakrabarty, A. M. 1976. Plasmids in *Pseudomonas*. *Annu. Rev. Genet.* **10:**7–21.

May, T. B., D. Shinabarger, R. Maharaj, J. Kato, L. Chu, J. D. DeVault, S. R. Choudhury, N. Zielinski, A. Berry, R. K. Rothmel, T. K. Misra, and A. M. Chakrabarty. 1991. Alginate synthesis by *Pseudomonas aeruginosa*: A key pathogenic factor in chronic pulmonary infections of cystic fibrosis patients. *Clin. Microbiol. Rev.* **4:**191–206.

DAVID A. HOPWOOD

A Love Affair with *Streptomyces* Genetics

Looking back to a slower era, I was born in the English Midlands, but I spent most of my early years near Portsmouth on the south coast of England. We lived on the edge of a small town, only a few minutes' walk to the countryside. This is where my mother, who loved nature, would take my older sister and me on country walks (in the earliest years I was in a stroller!) on most afternoons, and my mother's infectious enthusiasm for plants and animals soon rubbed off on us. I was only six when World War II broke out. In 1940, my sister and I were sent away to a boarding school in the north of England to escape the predicted bombing of Portsmouth, but this adventurous school—headed by Kenneth Barnes, a disciple of A. S. Neal of Summerfield fame and a committed exponent of free expression—turned out to be fraught with hazards of its own for the very young. So, back home we went after only three months. It was in the middle of the school year, so my mother, who was a qualified teacher, undertook to teach my sister and me at home until we reached secondary school age. Many of the "lessons" involved nature study, with regular trips to the still largely unspoiled countryside, with its almost total wartime absence of cars. For us it was rather idyllic, and so whenever I thought about it, a career involving plants or animals seemed the obvious goal. This was reinforced when my sister talked the local farmer into letting us give a hand on the farm on weekends and school holidays. So we helped with the harvest, rounded up the cows from the fields and milked them by hand, mucked out the cowsheds, and generally enjoyed the rural life.

At the end of the war, when I was 12, my father, who was in the Inland Revenue (Internal Revenue Service), was transferred to the industrial city of Manchester, and schoolwork and activities were taking up most of my time, so the farming lapsed. It must have been soon after that when I saw a documentary at the local cinema about the Agricultural Advisory Service. It portrayed a day in the life of a scientist who visited farms and collected samples, perhaps to check on soil fertility or a plant disease, and took them back to the lab to sort the problem out. By then I was really keen on science, especially because the school had an outstanding science staff, so I planned a career that combined the open air life on the farm with scientific research.

On leaving school, a degree in agriculture seemed the obvious next step, but agricultural departments in universities were closing; practical farming would be taught in colleges of agriculture, and the underpinning science would be taught in the pure science departments of universities. The advice was to earn a degree in pure science and then take a postgraduate diploma in agricultural studies.

I went to Cambridge University in October 1951 and took courses in botany, zoology, biochemistry, and geology in the first two years. I was won over by laboratory science and forgot about the idea of an outdoor life! I found genetics the most exciting topic, partly because, although we had covered a wide range of biology in high school, genetics was a total blank. This was the main motivation for choosing botany as my final year option in the academic year 1953 to 1954, the year in which Watson and Crick reported the structure of DNA. In the Cambridge Botany School, Harold Whitehouse and Lewis Frost were studying "biochemical genetics" using the fungus *Neurospora crassa*. They farsightedly emphasized the promise of the new field of microbial genetics for the future understanding of gene structure and function.

One of the Ph.D. topics on offer was the actinomycete *Streptomyces*. Here was a group of microbes apparently "intermediate between bacteria and fungi," so they could not fail to be interesting genetically, given the clear differences that were emerging between genetic phenomena in the few bacteria so far studied and in fungi and higher organisms. In the pneumococcus, *Escherichia coli* and *Salmonella typhimurium*, incomplete genomes were transferred from donor to recipient strains by one of three bizarre processes (transformation, conjugation, and transduction) to yield incomplete zygotes, whereas fungi and higher organisms had life cycles, including a complete diploid stage and meiosis.

To get me started, Harold Whitehouse lent me a book by Selman Waksman, the pioneer of actinomycete biology and discoverer of streptomycin. Lewis Frost gave me some *Streptomyces* cultures. I streaked them out and chose one that produced a striking blue pigment. I set about isolating auxotrophic mutants from this *Streptomyces coelicolor* strain in order to look for genetic recombination in the way that Joshua Lederberg and Edward Tatum had done ten years earlier in *E. coli*. I could grow pairs of mutants together and select rare prototrophs from the progeny spores, and in crosses of two doubly marked strains, nonselected markers from each parent segregated amongst the progeny, indicating a genuine process of gene reassortment rather than simply some kind of heterozygote formation. It seemed that I was in business, but meanwhile the true phylogenetic relationships of *Streptomyces* needed looking into.

Early in the development of microbiology in the 1870s, there were several descriptions of novel organisms that are now included in the actinomycetes, including the leprosy bacillus and the causal agent of "lumpy jaw" in cattle. It was probably the naming of the latter organism as *Actinomyces* ("ray fungus"), because of the tendency of the elongated, branching cells to grow radially from the center of the lesion, that confused later microbiologists about the taxonomic position of the *Actinomycetales*. By the early 1950s, few microbiologists regarded the actinomycetes as fungi, but many still thought of them as intermediate between fungi and bacteria. For *Streptomyces* there were biochemical pointers to a bacterial affinity, but their cellular architecture was unclear. In a rewarding collaboration with Audrey Glauert at the Strangeways Laboratory in Cambridge, we showed that *S. coelicolor* lacked a nuclear membrane and so was, by definition, a prokaryote. Nevertheless, it was clearly different from the simple, rod-shaped bacteria that had hitherto been studied genetically, so it might well reveal novel genetic phenomena.

While still a Ph.D. student, I was appointed to the lowest rung of the teaching ladder in the Cambridge Botany School in 1957, and soon I became a research fellow, first of Magdalene College and then of St. John's. The university job was to organize practical courses and demonstrations, while my college duties involved teaching groups of two or three students and was especially rewarding for the small-group interactions that it allowed.

I took my *Streptomyces* research to Glasgow in 1961, when I was appointed to a lectureship in Guido Pontecorvo's Genetics Department. Ponte was just switching from his groundbreaking research in *Aspergillus* to pioneer human somatic cell genetics, and I supposed that he would suggest that I also switch fields. In response to his question, "Are you going to carry on with *Streptomyces*?" I casually replied, "For the time being." He asked rhetorically, "Why not forever?" This might have been a throwaway remark, but Ponte was not prone to them. In any event, I took it as a seal of approval.

Life in Ponte's department was constantly stimulating, not only because of his inimitable iconoclasm and lively scientific imagination but also for the stream of key figures in the emerging field of molecular biology who visited: Salvador Luria, Renato Dulbecco, Francis Crick, James Watson, Maurice Fox, Bill Hayes, Enrico Calef, Frank Stahl, Werner Maas. . . . I was on the edge of bacterial genetics—*E. coli*, *Salmonella*, and the pneumococcus occupied center stage—but these giants took a genuine interest in what I was doing.

This was the "golden age" of bacterial and phage genetics, characterized by a rare collaborative spirit, infectious enthusiasm, and lack of concern about confidentiality and patent rights that look rosy in hindsight. Werner Maas invited my family and me to New York City for a six months' sabbatical in 1967 to teach his postgraduate course in microbial genetics. This was a truly broadening experience, both scientifically and culturally. We were living in Greenwich Village during a time of dramatic social change, including demonstrations against the Vietnam war—a period of history that passed almost before we knew it.

Back in Glasgow at the end of 1967, I had itchy feet, so when the job of John Innes Professor of Genetics at the University of East Anglia in Norwich and Head of the Genetics Department at the John Innes Institute opened up, I applied and was offered the job. It was rough going initially because the relationship between the university and the rump of the once-mighty John Innes Institute, transferred from its previous rural location to Norwich, still had to be worked out.

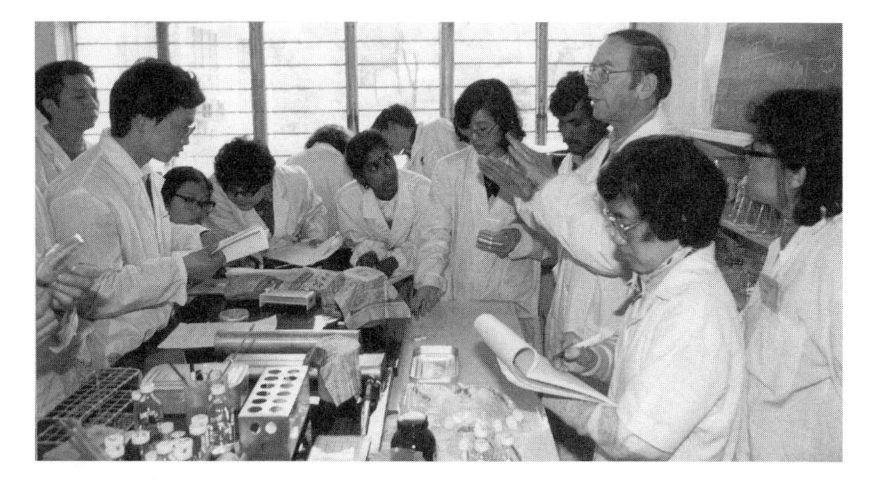

Teaching a UNIDO-sponsored course on Streptomyces *genetics in Wuhan, China, April 1989.*

Gradually a productive symbiosis between university and institute developed, and this synergy between student contact at the University of East Anglia and research at the John Innes has provided a wonderfully stimulating career. Soon I was able to hire some excellent students and colleagues, and gradually *Streptomyces* genetics entered its own golden age, the benefits of which we continue to reap.

My 45-year love affair with *Streptomyces* genetics has gone through three phases, which I dub the *in vivo*, *in vitro*, and *in silico* years, occupying roughly the end of the 1950s along with the 1960s and 1970s, the 1980s, and the 1990s, respectively. Building on my discovery of genetic recombination in *S. coelicolor* (which turned out not to be a lone achievement because several other groups had started on a similar hunt at the same time, notably Giuseppe Sermonti and his wife, Isabella Spada-Sermonti, in Rome), I devised a novel way to build a linkage map in a bacterium such as *Streptomyces* in which all recombinants have to be selected from a vast excess of asexually produced progeny. Using it, I showed by 1965 that *S. coelicolor*—like the two other bacterial examples available—had a single, circular linkage group, thereby helping to establish it as the bacterial paradigm. The paradigm has held up for the bacterial chromosome, but ironically, 30 years later we showed that the chromosome of *S. coelicolor* (though not its genetic linkage map) is actually linear!

The idea that plasmids controlled various characteristics in *Streptomyces* had arisen through the finding that melanin production, or the ability to sporulate, was often readily lost. We tried hard to prove the presence of plasmids and were rewarded by the discovery of SCP1. This plasmid became the first—and still the only well-proved—example of a plasmid that carries a set of genes for biosynthesis of an antibiotic (methylenomycin). SCP1 was defined genetically: It was totally refractory to physical isolation, a circumstance that was only explained 10 years later, when Haruyasu Kinashi showed it to be a large (350 kb), *linear* molecule. It could thus be separated from chromosomal DNA only by the later-developed tool of pulsed-field gel electrophoresis.

In the 1970s, we added artificial protoplast fusion to natural, plasmid-mediated conjugation as a means of doing *Streptomyces* genetics and then, much more important, the ability to transform protoplasts with plasmid or bacteriophage DNA, thus ushering in the *in vitro* years of gene cloning, which really put *Streptomyces* genetics on the map. Soon, numerous antibiotic resistance genes had been isolated, and basic aspects of gene expression worked out. Nevertheless, it was the cloning and analysis of antibiotic biosynthetic genes that gave me the most thrills. We now could manipulate the production of antibiotics by streptomycetes, which had enormous potential importance for the emergence of biotechnology in the pharmaceutical industry.

When I chose *Streptomyces* as a topic for genetic study, it was irrelevant for me (though not for some of the other pioneers) that the actinomycetes were becoming known as the world's greatest producers of antibiotics, soon to include such workhorse antibacterials as tetracycline and erythromycin, as well as more esoteric compounds like the anticancer agent Adriamycin and the antiparasitic agent avermectin, constituting a multibillion-dollar industry.

By the 1980s, my view had changed, with the realization that the blue pigment of *S. coelicolor* is an antibiotic (actinorhodin), the color of which gave antibiotic genetics a big helping hand. We isolated mutants by their loss of pigment or its alteration, cloned the genes by restoring color to a blocked mutant, and, in 1985, reported the first "hybrid" antibiotics made by genetic engineering, again spotted initially by a changed pigmentation. I was especially elated by this achievement because it involved an excellent international collaboration with leaders in natural product chemistry in Japan and in the United States.

International friendships have been among the most satisfying aspects of my career. Over the years, scientists from many countries have learned the art of *Streptomyces* genetics by coming to the John Innes Institute as Ph.D. students or postdocs or attending one of the EMBO training courses in Norwich, the teaching of which was tremendously stimulating to my colleagues and me between 1983 and 1990. The result has been an extensive international network of friends and collaborators.

The 1990s saw an explosion of *in silico* genetics in *Streptomyces*, as in genetics as a whole. It was the major facilitator for understanding the genetic "programming" of polyketide synthases, those amazingly versatile multifunctional enzymes that put together the huge range of molecules built from acetate, malonate, and other simple carboxylic acids, with variations in chain length, level of oxidation at alternate carbons, chirality at asymmetric centers, and ring formation, which gives this family of natural products their unique biological activities and medical or agricultural importance. Out of this has come the new field of "combinational biosynthesis" of "unnatural natural products," which may finally see the long years of academic *Streptomyces* genetics earn its commercial keep within a year or two.

Meanwhile the *Streptomyces* chromosome continues to amaze. In a rewarding collaboration, Carton Chen from Taipei and my career-long scientific partner, Helen Kieser, showed it to be linear but capable of converting to a circular form—a real novelty. Helen made a physical map of the *S. coelicolor* chromosome, and she, Matthias Redenbach, and friends put detail onto the combined genetic/physical map with the construction of an ordered library of cosmid clones covering the whole chromosome in about 320 bite-sized pieces. This in turn provided the basis for a whole-genome sequencing project started at the Sanger Centre in Cambridge in August 1997 and due to be completed by late 2000. Already the sequence of the *S. coelicolor* chromosome—at 8 Mb and over 7,000 genes the largest bacterial genome currently in play—is revealing genetic novelties on a daily basis. It is most satisfying to have reached this stage in my career. It will ensure my continuing fascination with *Streptomyces* genetics and permit the greater biotechnological productivity of streptomycetes in the years to come.

DAVID A. HOPWOOD was born in Kinver, Staffordshire, England. He received a bachelor's degree in 1954 from Cambridge University, a doctoral degree in 1958 from the same institution, and a D.Sc. in 1974 from the University of Glasgow. He also has received an honorary doctorate of science from the Eidgenössische Technische Hochschule, Zürich, in 1989; an honorary doctorate from the University of Manchester Institute of Science and Technology in Manchester, England, in 1990; and an honorary doctorate of science from the University of East Anglia, Norwich, in 1998. He was Assistant Lecturer in Botany at Cambridge from 1957 to 1961. He then was Lecturer in Genetics at the University of Glasgow from 1961 to 1968. He became John Innes Professor of Genetics at University of East Anglia in Norwich and Head of the Genetics Department of the John Innes Institute in 1968. In 1998 he was named Emeritus Professor of Genetics at the University of East Anglia and Emeritus Fellow at the John Innes Centre. He became a fellow of the Royal Society of London in 1979 and has received numerous awards, including the Medal of the Kitasato Institute, Tokyo, Japan, 1988; the Hoechst-Roussel Award, American Society for Microbiology, 1988; Honorary Member, Hungarian Academy of Science, 1990; Chiron Biotechnology Award, American Society for Microbiology, 1992; Knight Bachelor, 1994; Mendel Medal of the Czech Academy of Sciences, 1995; and Gabor Medal of the Royal Society, 1995.

The following papers are representative of his publications:

Redenbach, M., H. M. Kieser, D. Denapaite, A. Eichner, J. Cullum, H. Kinashi, and D. A. Hopwood. 1996. A set of ordered cosmids and a detailed genetic and physical map for the 8 Mb *Streptomyces coelicolor* A3(2) chromosome. *Mol. Microbiol.* **21:**77–96.

McDaniel, R., S. Ebert-Khosla, D. A. Hopwood, and C. Khosla. 1995. Rational design of aromatic polyketide natural products by recombinant assembly of enzymatic subunits. *Nature* **375:**549–554.

Lin, Y. S., H. M. Kieser, D. A. Hopwood, and C. W. Chen. 1993. The chromosomal DNA of *Streptomyces lividans* 66 is linear. *Mol. Microbiol.* **10:**923–933.

Hopwood, D. A., F. Malpartida, H. M. Kieser, H. Ikeda, J. Duncan, I. Fujii, B. A. Rudd, H. G. Floss, and S. Omura. 1985. Production of "hybrid" antibiotics by genetic engineering. *Nature* **314:**642–644.

Malpartida, F., and D. A. Hopwood. 1984. Molecular cloning of the whole biosynthetic pathway of a *Streptomyces* antibiotic and its expression in a heterologous host. *Nature* **309:**462–464.

Hopwood, D. A., H. M. Wright, M. J. Bibb, and S. N. Cohen. 1977. Genetic recombination through protoplast fusion in *Streptomyces*. *Nature* **268:**171–174.

Hopwood, D. A. 1959. Linkage and the mechanism of recombination in *Streptomyces coelicolor*. *N.Y. Acad. Sci.* **81:**887–898.

Section Nine

OUT OF NATURE

"Nature is to be found in her entirety nowhere more than in her smallest creatures." Such was the view of Pliny, the natural historian who died in the year 79 while studying the eruption of Mt. Vesuvius. Although bacteria, admittedly, are not warm, fuzzy animals, the global web of life is balanced on these unseen microbes, most of which remain to be discovered. The world is filled with a sea of microbes swimming in the oceans and soils, which are their arenas for the endless chemical transformations they perform. So as some scientists looked at microbes in fermenters and diseases, others pursued these invisible creatures in their natural ecological habitats. Exploring the microbial processing of minerals in soils in his home city of Delft at the end of the nineteenth century, Martinius Beijerinck continued the heritage of van Leeuwenhoek by studying the roles of microbes in nature, a scientific lineage continued through the twentieth century by Albert Kluyver and C. B. van Niel. "Scientific exploration of microbes in the laboratory has moved back into an ecological world—a world of unusual microbes living in strange, and often hostile, places—a world in which most microbes are still unknown."

As seen by the naturalist and popular writer Stephen Jay Gould in *The Panda's Thumb*, "The world of a bacterium . . . is so unlike our own that we must abandon all our certainties about the way things are and start from scratch. . . . The world is full of signals we do not perceive. Tiny creatures live in a different world of unfamiliar forces . . . what an imperceptive lot we are." Not surprising, then, looking back at the 1898 meeting of the Society of American Naturalists, that one of the participants would remark: "You bacteriologists seem to be wandering around like lost souls—why don't you have a society of your own?" And so about forty bacteriologists in the United States and Canada marched out and formed a Society of American Bacteriology, which later would become the American Society for Microbiology with more than 42,000 microbiologists as it reached its centennial year and celebrated its 100th annual meeting.

It was not the nature of the organisms but, rather, the nature of scientific inquiry that forced microbiology to emerge from the society of naturalists. Pasteur had led the charge to hypothesis-driven science, and Koch had provided the methods for experimentation. The microbe hunters that followed could not just view the beauty of the microbial world.

> If microorganisms . . . excite the curiosity of scientists . . . it is because . . . the simplest of organized beings present striking analogies with more highly developed living organisms, but unlike the latter, they can be used with marvelous facility for the most delicate scientific investigations.
>
> —Lechevalier and Solotorovsky, *Three Centuries of Microbiology*

All Because of a Butterfly

One of my earliest memories is seeing a European swallowtail butterfly in the woods of Germany. I was five years old. Every weekend the children of our village, accompanied by adults, took a hike through the meadows and forest.

When I came to the United States in 1938, at the age of eight, my mother, father, sister Helen, and I settled in Grand Forks, North Dakota, where our relatives were among the first Europeans to arrive, in 1880. Mother and Father, having been butchers in Germany, opened a small neighborhood grocery store, which they maintained for 30 years.

My interest in butterflies continued and grew. I acquired a large collection of local and other American species, as well as many tropical and Asian ones. I remember the "cat lady" of our neighborhood telling me how terrible it is to kill a butterfly, but I remained unconvinced until the age of 35. (Now my collection sits in our attic, eaten up by beetle larvae; it is waiting for me to throw it out.)

Ultimately, not only butterflies but also all other insects—and all other animals and flowers—became the center of my interests, second to my wife, Hilde, and children, David and Jean. Originally the fascination with living things was an aesthetic one, but this evolved into a curiosity about their behavior. Eventually, after I had studied organic chemistry, biochemistry, and genetics, the interest turned into how the behavior of all creatures, including humans, is accomplished.

For example, what makes the monarch butterfly choose milkweeds for depositing its eggs? And what makes the monarch caterpillar stay on the milkweed to feed until maturity? I thought it must be explained by the butterfly's smelling of the volatile chemicals given off by the milkweed and the caterpillar's tasting probably other chemicals. What are these chemicals? How are they sensed? How are they acted on? In short, what are the mechanisms of all this? And how do those mechanisms relate to the behavior of organisms in general?

After earning a bachelor's degree at Harvard University in 1952 and a Ph.D. with Henry Lardy at the University of Wisconsin in Madison in 1957, I took a course in microbiology with C. B. van Niel at the Hopkins Marine Station in Pacific Grove, California. There I found a huge library dating back into the eighteenth century. I learned that Wilhelm Pfeffer, a famous German botanist, in 1880 had used motile bacteria to study attraction and repulsion by various plant and animal extracts and by the few chemicals known at that time. I decided that was the way to go! By now so very much more was known about the chemistry of thousands of natural compounds, and by now the biochemistry and genetics of a commonly used bacterium (*Escherichia coli*) were familiar and available for use. I decided to study the behavior of bacteria and then ultimately to broaden out to the behavior of all other organisms.

To acquaint myself further with biochemistry and genetics, I did postdoctoral studies with Arthur Kornberg and Dale Kaiser at Washington University and then at Stanford University. Afterward, in 1960, I accepted an offer from the Departments of Biochemistry and Genetics at the University of Wisconsin to become a professor, and I have been in Madison ever since.

The past forty years I have been working on the behavior of bacteria—the genetics and the biochemistry of it. How do bacteria sense the world around them? Do they do it the way we do, with sensory receptors for taste and smell and vision and touch and temperature? Or instead do they lack sensory receptors, and do these stimuli simply speed up or slow down the various changes that take place inside them so that they congregate where conditions are good and avoid places where toxic things occur?

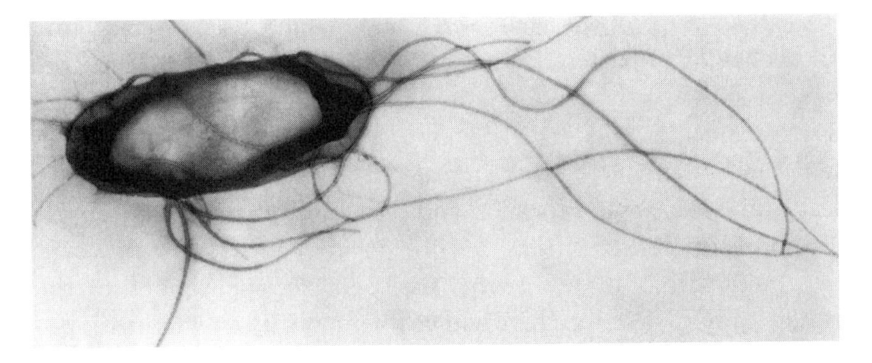

A bacterium with flagella.

I showed that bacteria indeed do have sensory receptors. Some chemicals are not attractants for motile bacteria, even though the bacteria can use them perfectly well, simply because sensory receptors are not there for those chemicals. Other chemicals are excellent attractants whether or not the chemicals get used, and in some cases even if the chemicals are unable to get inside the bacteria (like something we smell without its getting inside of us), because sensory receptors are present for such chemicals on the outside of the bacteria.

Then we isolated mutants that can't respond to this or that attractant because of missing this or that receptor (like people who can't smell or taste certain things). Next we isolated other mutants that are not attracted or repelled by anything at all, though fully motile; they are missing the pathway from sensory receptors to the organs of locomotion—the flagella (like a person defective in the part of the nervous system that carries information from the nose or tongue to the legs).

The questions now became: What is the nature of the bacterial sensory receptors, and how is the sensed information carried from the receptors to the flagella? We then identified the sensory receptors as structures specialized for sensing an attractant or a repellent (methyl-accepting chemotaxis proteins). Also, we identified some of the components that carry the information from the sensory receptors to the flagella (chemotaxis, or Che, proteins).

Soon this research subject attracted many biochemists and geneticists and molecular biologists and neurobiologists and mathematicians and physicists from all over the world. These people carried the project much, much farther. Consequently we have today a very good picture of how it all happens, though the story is still far from complete.

Does anything of what has been found out about sensory reception and behavior in bacteria carry over to higher organisms? Much the same story has now been found in higher single-celled creatures such as yeast and algae and fungi, and also in plants. But it has not yet been found in butterflies, in other animals, or in people. We are trying to learn now if it applies to smell and taste and other ways of sensing chemicals in these organisms.

I predict that in the twenty-first century, applications of the research I have described here will be found. If the system turns out to be universal in occurrence, then the work may well lead to the understanding and treatment of diseases involving defects in sensory receptors, nerves, muscles, and behavior. Even if it does not, the results will be useful for fighting diseases caused by pathogens that do have this system of sensory reception.

What started out as a fascination with the beauty of a butterfly has turned into a lifelong pursuit of the mechanisms of behavior in all living things—a pursuit that promises practical applications but has been driven by the joy of discovery.

> The scientist does not study nature because it is useful to do so. He studies it because he takes pleasure in it; and he takes pleasure in it because it is beautiful. If nature were not beautiful, it would not be worth knowing and life would not be worth living. . . . I mean the intimate beauty which comes from the harmonious order of its parts and which a pure intelligence can grasp.
>
> —Jules Henry Poincaré (1854–1912)

JULIUS ADLER was born in Edelfingen, Germany, in 1930. He received a bachelor's degree from Harvard University in 1952. He earned a doctoral degree in biochemistry in 1957 from the University of Wisconsin. He did postdoctoral studies at Washington University and then at Stanford University. In 1960 he joined the faculty of the Departments of Biochemistry and Genetics at the University of Wisconsin. He is now Professor Emeritus. He is a member of the National Academy of Sciences and has received the Waksman Award and the Otto Warburg Medal for his research on the biochemistry and genetics of the behavior of microorganisms. He also received the Abbott-American Society for Microbiology Lifetime Achievement

Award and the American Society for Biochemistry and Molecular Biology's William C. Rose Award.

The following papers are representative of his publications:

Tisa, L. S., and J. Adler. 1995. Chemotactic properties of *Escherichia coli* mutants having abnormal Ca^{2+} content. *J. Bacteriol.* **177:**7112–7118.

Lake, E. M., H. Jiang, F. R. Blattner, and J. Adler. 1995. Analogues of aspartate and glutamate active at synapses are attractants for *Escherichia coli*. *Cell. Molec. Neurobiol.* **15:**283–288.

Li, C., and J. Adler. 1993. *Escherichia coli* shows two types of behavioral responses to osmotic upshift. *J. Bacteriol.* **175:**2564–2567.

Adler, J. 1987. How motile bacteria are attracted and repelled by chemicals: An approach to neurobiology. *Biol. Chem. Hoppe-Seyler* **368:**163–173.

Eisenbach, M. and J. Adler. 1981. Bacterial cell envelopes with functional flagella. *J. Biol. Chem.* **256:**8807–8841.

Goy, M. F., M. S. Springer, and J. Adler. 1977. Sensory transduction in *Escherichia coli*: Role of a protein methylation reaction in sensory adaptation. *Proc. Natl. Acad. Sci. USA* **74:**4964–4968.

Springer, M. S., M. F. Goy, and J. Adler. 1977. Sensory transduction in *Escherichia coli*: Two complementary pathways of information processing that involve methylated proteins. *Proc. Natl. Acad. Sci. USA* **74:**3312–3316.

Adler, J. 1976. The sensing of chemicals by bacteria. *Sci. Am.* **234:**40–47.

Adler, J. 1975. Chemotaxis in bacteria. *Annu. Rev. Biochem.* **44:**341–356.

Kort, E. N., M. F. Goy, S. H. Larsen, and J. Adler. 1975. Methylation of a membrane protein involved in bacterial chemotoxis. *Proc. Natl. Acad. Sci. USA* **72:**3939–3943.

Adler, J., G. L. Hazelbauer, and M. M. Dahl. 1974. Chemotaxis towards sugars in *Escherichia coli*. *J. Bacteriol.* **115:**824–847.

DePamphilis, M. L., and J. Adler. 1971. Fine structure and isolation of the hook-basal body complex of flagella from *Escherichia coli* and *Bacillus subtilis*. *J. Bacteriol.* **105:**384–395.

Armstrong, J. B., and J. Adler. 1969. Location of genes for motility and chemotaxis on the *Escherichia coli* genetic map. *J. Bacteriol.* **97:**156–161.

Adler, J. 1969. Chemoreceptors in bacteria. *Science* **166:**1588–1597.

It Came from Chicken Water

Our high school sophomore biology teacher asked the class to bring in a water sample from home to examine under the microscope. He suggested we take water from where the cows, chickens, or pets drank, so my father's chicken water was a natural source for me to sample. One look at my chicken water under the microscope and the normally reserved biology teacher became completely excited; the sample was loaded with microbes! Amoebae, paramecia, algae—many busy, motile bacteria and oddly shaped characters floated by our eyes. The whole class was delighted, and I was charged with bringing in a sample of that chicken water for every biology class the rest of the time I was in high school. The excitement of this first exposure to microbiology has always stayed with me.

From the time I was eight years old, it had been taken for granted in my family that after high school I would enroll in the nearest campus of the University of California, located in Davis, but my future as a bacteriologist was still far from anyone's mind. UC Davis was primarily an agricultural school at that time, but it also had a school of veterinary medicine and ran its own creamery. My uncle had a dairy close to UC Davis and sold his milk to the creamery. The UC Davis veterinarians and their students who came to treat my uncle's sick cows fascinated my cousins and me. I knew about many microbial diseases long before I knew about microbes; outbreaks of mastitis, anthrax, pink eye, brucellosis, and so on were fairly common occurrences on the farm. I was very impressed by the fact that when a cow died of anthrax, my uncle had to burn her on the spot and flame the surrounding grass as well.

The closest profession to microbiology that my high school career counselor, a nurse, could recommend was being a medical laboratory technician. With this goal in mind, I asked, on acceptance at UC Davis, to be put into the suggested training program as a biochemistry major. In fact, UC Davis had no undergraduate biochemistry major, so I was assigned to bacteriology instead. This turned out to be a wonderful and fortuitous match, despite the apparently arbitrary way the decision was made. The Bacteriology Department at UC Davis at that time had a truly remarkable and renowned faculty. For instance, John Ingraham, the eminent microbial physiologist; Robert Hungate, who brought anaerobic microbiology into its own; and Mortimer P. Starr, an internationally renowned expert in microbial taxonomy were all on the faculty at that time. Starr's classes were adventures in sampling—from vinegar factories and brewing houses to salt ponds and San Francisco Bay mud. I specialized in cow pies from my uncle's dairy and purified what was then the largest known bacterium, *Caryophanon*.

Monty Reynolds taught the introductory microbiology course and was one of the most memorable professors I have ever had. An extremely hyperkinetic type, he literally jumped around the lectern. He also showed a lot of unforgettable movies, many obtained from Army archives. We saw the consequences of a rabid wolf biting villagers in a remote area of Iran and a film of a UC Berkeley student who had accidentally pricked himself with a rose thorn and become infected with tetanus (the student allowed the course of his infection and recovery to be filmed). The images are still vivid in my mind's eye.

My introduction to microbiological research began in the summer between my junior and senior years. I was the lucky recipient of a National Science Foundation undergraduate research fellowship and was picked to work in John Ingraham's laboratory. He put me to work studying the phenomenon of cold shock—what goes on when bacterial cells are subjected to a sudden large drop in growth temperature. I was fascinated by the topic and continued to pursue it until graduation. By this time, I was hooked on research and a bit confused about the medical lab tech option. Dr. Ingraham rescued me by offering a research technician's job in his lab. I was delighted to accept. My role was to try out Ingraham's novel and exciting ideas. If they worked, the project was given to a graduate student or postdoctoral fellow; if not, the idea was dropped. I quickly learned that I didn't have enough information to be a major contributor to the new ideas. It was clear that I needed more class work and training—my motivation for pursuing a master's degree in microbiology at UC Davis.

During this period, I married Craig Squires, another technician in the Bacteriology Department. Craig was anxious to move to UC Santa Barbara and work for Ellis Englesberg, studying a novel positive regulatory system—the arabinose operon. The UCSB Biology Department welcomed us and gave me several options: work as a technician, run the microbiology teaching labs, or become a Ph.D. student. By that time, I was totally hooked on research but still thought I didn't know enough to be an insightful and independent researcher, so I opted for the doctoral program. Again, I was fortunate to receive superb training. My degree in biochemistry and molecular biology was completed under the supervision of an outstanding mentor, Nancy Lee. I have always tried to emulate her wonderfully rigorous and thoughtful approach to science. By the time I received my degree, I was confident that I could purify and analyze virtually any protein.

For postdoctoral training I wanted to work on one of the major regulatory puzzles of the time: how amino acid biosynthetic operons are regulated. I knew that to stay in research and be successful, I would need the same kind of mentoring, confidence in my abilities, and networking help I had received from John Ingraham, Nancy Lee, and many of the other faculty at UC Davis and UC Santa Barbara. Somehow I knew enough to seek out Charles Yanofsky at Stanford University and ask to work on the tryptophan operon. Yanofsky took on Craig as a research associate as well, and this time was the beginning of an exciting period during which the group collectively discovered the novel regulatory phenomenon of transcription attenuation.

Again, I was very fortunate to have a superb mentor. When the time came for me to leave Stanford, I agonized over what my independent research project should be when I took my first faculty position. After several months, I settled on the mysterious regulation of ribosome biosynthesis. Building on the success of Naomi Franklin in making fusions between phage promoters and tryptophan biosynthetic genes, I decided that the gene fusion approach would provide a unique handle on studying ribosomal gene expression. The only problem was that I wasn't yet schooled in the newly created recombinant DNA technology. Yanofsky allowed me to spend the last four months of my postdoctoral training period in Herb Boyer's lab at UC San Francisco. There, Boyer and Pat Green taught me how to do endonuclease restriction digests, ligations, and transformations and to run agarose gels; in short, they taught me the fundamentals of the emerging recombinant DNA technology.

I have worked on the regulation of ribosomal RNA (rRNA) synthesis ever since that time and feel very fortunate to have had excellent students, postdoctoral fellows, and associates to share this project. We have applied genetics, biochemistry, molecular biology, and physiology—the patchwork of my training and interests—to our studies of ribosome synthesis. We have studied transcription regulation, using gene fusions to characterize a special feature of rRNA synthesis—transcription antitermination. We have also studied cell physiology and genetics as they relate to copies of ribosomal genes. We were successful in constructing a strain of *Escherichia coli* with no intact chromosomal rRNA operons. The cell's only source of ribosomal genes is a plasmid. By exchanging *coli* sequences on the plasmid with those of other microbes, we are able to manipulate a cell's ribosome content in many unexpected ways and thereby pose novel questions about the cell's translation machinery.

I remain as enthusiastic and excited about microbiological research today as when I started more than 30 years ago. My advice to students and new researchers is always to try to pursue what really interests and excites you; there is no substitute for enthusiasm in research.

CATHY SQUIRES was born in Sacramento, California, in 1941. She received her bachelor's and master's degrees from the University of California at Davis in 1963 and 1967, respectively. Her doctoral degree is in molecular biology and biochemistry from the University of California at Santa Barbara. She did postdoctoral studies at Stanford University. She is now Professor and chair in the Department of Molecular Biology and Microbiology at Tufts University School of Medicine. Her research focuses on the mechanisms of transcription and translation in bacteria.

The following papers are representative of her publications:

Voulgaris, J., S. French, R. Gourse, C. Squires, and C. L. Squires. 1999. Increased *rrn* gene dosage causes intermittent transcription of rRNA in *Escherichia coli*. *J. Bacteriol* **181**:4170–4175.

O'Conner, M., T. Asai, C. L. Squires, and A. Dalhberg. 1999. Enhancement of translation by the downstream box does not involve base pairing of mRNA with the penultimate stem sequence of 16S rRNA. *Proc. Natl. Acad. Sci. USA* **96**:8973–8978.

Asai, T., D. Zaporojets, C. Squires, and C. L. Squires. 1999. An *E. coli* strain with all chromosomal rRNA operons inactivated: Complete exchange of rRNA genes between bacteria. *Proc. Natl. Acad. Sci. USA* **96:**1971–1976.

Condon, C., S. French, C. Squires, and C. L. Squires. 1993. Depletion of functional ribosomal RNA operons in *Escherichia coli* causes increased expression of the remaining intact copies. *EMBO J.* **12:**4305–4315.

Condon, C., J. Philips, Z.-Y. Fu, C. Squires, and C. L. Squires. 1992. Comparison of the expression of the seven ribosomal RNA operons in *Escherichia coli. EMBO J.* **11:**4175–4185.

Berg, K., C. L. Squires, and C. Squires. 1989. Ribosomal RNA operon antitermination: Function of leader and spacer region box B-box A sequences and their conservation in diverse microorganisms. *J. Mol. Biol.* **209:**345–358.

Li, S., C. L. Squires, and C. Squires. 1984. Antitermination of *Escherichia coli* ribosomal RNA transcription is caused by a control region segment containing lambda *nut*-like sequences. *Cell* **38:**851–860.

Barry, G., C. Squires, and C. L. Squires. 1980. Attenuation and processing of the messenger RNA from the *rp1JL-rpoBC* transcription unit of *Escherichia coli*. *Proc. Natl. Acad. Sci. USA* **77:**3331–3335.

Bertand, K., L. Korn, F. Lee, T. Platt, C. L. Squires, C. Squires, and C. Yanofsky. 1975. New features of the regulation of the tryptophan operon. *Science* **189:**22–26.

Squires, C. L., F. Lee, and C. Yanofsky. 1975. Interaction of the *trp* repressor and RNA polymerase with the *trp* operon. *J. Mol. Biol.* **92:**93–111.

Captivating Methanogens

I was a twenty-five-year-old graduate student when I first saw living bacteria swimming in a wet mount. I was instantly captivated. I thought it was fantastic that bacterial cells could be inoculated into a sterile medium in the afternoon and the very next day one could have a fully grown culture. You could set up an experiment today and read the results tomorrow. This was different; this was a brand new biology to me. Throughout my schooling I had been interested in biology and had even majored in biology in college. Yet animals and plants were so complex and took so long to grow that I remained uncommitted. A course taught by W. G. Hutchinson at the University of Pennsylvania turned me on to bacteriology. So when Professor Hutchinson offered me a position as his teaching assistant in bacteriology, I was elated; I had found what I wanted to study—bacteriology.

Perhaps the reason I became fascinated with living bacteria was due to a poor background in microbiology. During a botany laboratory in college, we had been given some prepared slides of bacterial cells to examine under 400× magnification. One slide contained tiny red dots, the other tiny blue dots. No one seemed to understand that higher magnification and living cells are really needed to begin to reveal the great diversity of the microbial world.

As a graduate student, I was attracted to microbial physiology and metabolism because these were "moving areas" at the time. Techniques such as sonic oscillation had been developed for the rupture of bacterial cell walls; enzyme activity in cell extracts could also be studied. D. J. O'Kane accepted me as a master's student in researching what bacteria did—how they made a living, rather than what they were called or how they were classified.

I never had a formal course in biochemistry, and sometimes it shows. I worked on the enzyme hippuricase (from *Streptococcus*), a hydrolytic enzyme that cleaved a peptide-like bond and was considered of interest because the mechanism of peptide bond synthesis was unknown at the time. My initial attempts at research were rather painful to my professor, I'm afraid, especially when I didn't even know how to plot the data I generated. I needed to be spoon fed. Eventually, I found that doing the experiment on my own without letting my professor know was fun. When I thought I had established a scientific fact, we would talk. I liked this system; the thrill of discovery on one's own is the best motivating force.

My initial goal in entering graduate school had been to obtain a master's degree so that I could teach at a small college. But I soon abandoned that goal. I had been seduced by research and had become all too aware that adequate facilities for doing research were rarely found in a small college. What I really needed was a research degree—a Ph.D.—and a position at an institution that could provide the proper equipment for doing research. After receiving a master of science degree, I decided to take a year off from graduate school. Graduate students were paid $950 for nine months, so it was necessary to save enough money during that period to live during the summer. But I would soon return to the research laboratory.

What encouraged me to consider the possibility of an academic position in a research environment involved the preparation of my first manuscript from part of my thesis. I was apprehensive about submitting a manuscript to the *Journal of Biological Chemistry* because I lacked confidence in writing. However, I carefully patterned the manuscript in the style of the journal and gave it to my professor. He returned it and announced, "I think this is fine, Ralph; let's send it in." We did, and it was accepted! This was the first time anyone had ever expressed approval of

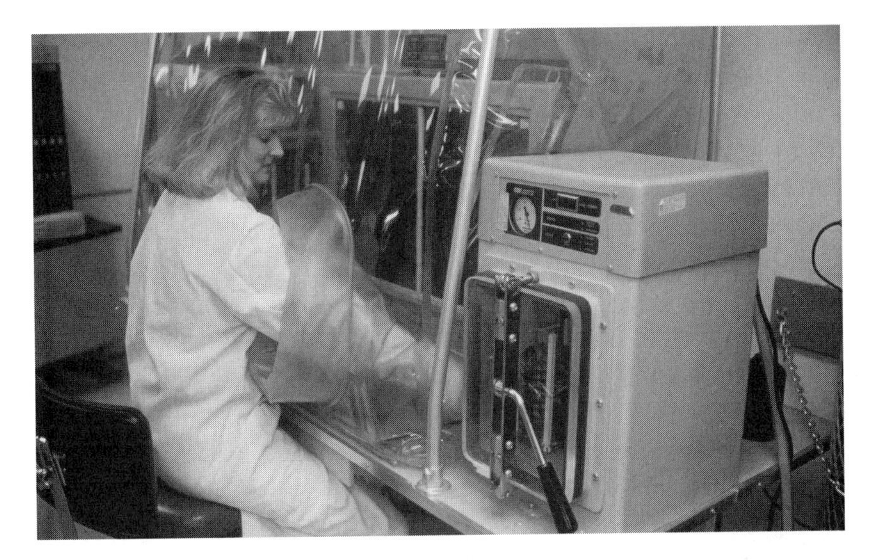

Inoculating cultures of strictly anaerobic microorganisms in an anaerobic glove box.

my writing and gave me confidence that perhaps I could become successful at scientific writing. I owe much to D. J. O'Kane, who made me appreciate the importance of hard data in nailing down a concept, as well as the importance of freedom in exploring and making discoveries. Later, I would use this same philosophy in running my own laboratory.

As a graduate student, I enjoyed laboratory teaching and was especially interested in unusual organisms. Because of my interests in microbial diversity, I was hired at the University of Illinois in 1953 as an instructor in microbiology. The Department of Bacteriology was an exciting place; with recent appointments of Halvorson, Spiegleman, Luria, Gunsalus, and Juni, the department was considered one of the best in the country, and I was fortunate to join it. A heavy teaching load didn't leave much time for research, and my program was rather slow in evolving. One day there was a knock on my office door. Professor Halvorson entered, sat down, and in a concerned manner said, "I just want to tell you one thing—you are paid to teach; you get promotions for doing research." He departed immediately, and I pondered these words of wisdom. They are as true today as they were then.

I never dreamed that bright students in my research program would lead me toward the discovery of not just one but seven new coenzymes! The first of these was ferredoxin, the electron carrier in the pyruvate clastic reaction—a new type of electron carrier protein. This discovery was made in collaboration with R. C. Valentine in my laboratory and Len Mortenson of the DuPont Company. In 1961, I began to study methanogenic bacteria, and these studies, over a twenty-year period, led to work on six new coenzymes of methanogenesis: coenzyme M, coenzyme F_{420}, tetrahydromethanopterin, coenzyme F_{430}, methanofuran, and coenzyme B. A major portion of my research career was devoted to enzyme-coenzyme relationships in the reduction of carbon dioxide to methane. Our studies on methanogenesis began with the culture known as *Methanobacillus omelianskii*, which, in collaboration with M. J. Wolin and M. P. Bryant, was shown to be a mixed culture of two organisms growing in a symbiotic relationship. One oxidized ethanol to acetate and hydrogen gas, while the other oxidized the hydrogen with reduction of carbon dioxide to methane. This was the discovery of interspecies hydrogen transfer in which the removal of hydrogen allows the anaerobic oxidation of an alcohol or fatty acid to become thermodynamically feasible—a first law of anaerobic microbial ecology.

The first new coenzyme of methanogenesis, coenzyme M (CoM), also was required as a vitamin by *Methanobacterium ruminantium*. Based on Kluyver's doctrine of comparative biochemistry, we expected the distribution of this new vitamin to be similar to lipoate or pantothenate. Assays of extracts from an exhaustive variety of organisms revealed that CoM was present solely in the methanogens. The doctrine of comparative biochemistry had let me down (!), but because we had developed the new procedure for growing methanogens, it was now possible to grow cells labeled with ^{32}P. We used this knowledge to begin collaborating with Carl Woese's lab to determine exactly how related these organisms were to other bacteria via comparative 16S RNA analysis. We couldn't believe the clear-cut results: Methanogens were not related to typical bacteria. Their unique metabolism and the unusual coenzymes supported this conclusion. This was the beginning of the "third form of life"—the archaea.

Over years of research I have formulated a few truisms, which I refer to as Wolfe's Laws of Thermodynamics. First Law: Unpublished data do not improve with age. Second Law: If you are first on the scene, it is easy to make discoveries. Third Law: The emotion generated in scientific discussion increases proportionally with the softness of the data being discussed. Fourth Law: If you join a parade, you become one of the marchers. Graduate students like to join highly visible parades, to be part of the current scene, not realizing that the number of marchers reduces their visibility. As an independent investigator, each should realize the importance of choosing a problem that isn't moving and move it.

In the summer of 1954, I became an observer in van Niel's famous course at Pacific Grove. This fantastic experience enabled me to start my own organisms course, which I would teach at the University of Illinois for 35 years. Through assignment of special problems, both the students and I learned something new about an unusual organism each time the course was given. I have had a continuing interest in unusual organisms: *Gallionella*, *Beggiatoa*, magnetic bacteria, photosynthetic bacteria, acetogens, and methanogens.

Throughout my academic life, I have more or less been driven in some manner "to go back to the lab." This common syndrome of persons in science needing to be in the lab could be simply a security blanket or a necessary component of scientific survival. I have thought that, for me, the latter was the case, coupled with a feeling that my presence somehow might encourage students. In operating my laboratory, my purpose has not been to play the role of the brilliant intellectual leader but rather to stay in the background and try to create an atmosphere in which students could develop into independent investigators. This, for me, is what it's all about.

I have found the research and teaching role of a professor to be both enjoyable and rewarding. I would choose again, without hesitation, the same career in microbiology if I were to start over. Microbial diversity is even more challenging today than it was four decades ago, because use of molecular probes has shown what many of us suspected—that only a tiny fraction of the microbial world has been cultivated. I believe that isolation and cultivation are prerequisites for a true understanding of microbes, and every microbiologist should have as a hobby the isolation of an unusual microbe from nature.

RALPH WOLFE received his doctoral degree from the University of Pennsylvania. He currently is Professor at the Center for Advanced Study at the University of Illinois in Urbana. He is a Fellow of the American Academy of Microbiology and the American Academy for the Advancement of Science, as well as a member of the National Academy of Sciences. He is Professor Emeritus of the University of Illinois. He was the 1999 recipient of the Procter & Gamble Award in Applied and Environmental Microbiology, largely for his contribution to the elucidation of the biological production of methane. His research unveiled the biochemistry of methane formation and helped establish principles of microbial ecology, physiology, biochemistry, and phylogeny that define a major portion of the broader field of applied and environmental microbiology.

The following papers are representative of his publications:

Metcalf, W. W., and R. S. Wolfe. 1998. Molecular genetic analysis of phosphite and hypophosphite oxidation by *Pseudomonas stutzeri* WM88. *J. Bacteriol.* **180:**5547–5558.

Metcalf, W. W., J. K. Zhang, E. Apolinario, K. R. Sowers, and R. S. Wolfe. 1997. A genetic system for Archaea of the genus *Methanosarcina*: liposome-mediated transformation and construction of shuttle vectors. *Proc. Natl. Acad. Sci. USA* **94:**2626–2631.

Metcalf, W. W., J. K. Zhang, X. Shi, and R. S. Wolfe. 1996. Molecular, genetic, and biochemical characterization of the *serC* gene of *Methanosarcina barkeri* Fusaro. *J. Bacteriol.* **178:**5797–5802.

Woo, G. J., A. Wasserfallen, and R. S. Wolfe. 1993. Methyl viologen hydrogenase II, a new member of the hydrogenase family from *Methanobacterium thermoautotrophicum* delta H. *J. Bacteriol.* **175:**5970–5977.

DiMarco, A. A., T. A. Bobik, and R. S. Wolfe. 1990. Unusual coenzymes of methanogenesis. *Annu. Rev. Biochem.* **59:**355–394.

Rouviere, P. E., and R. S. Wolfe. 1988. Novel biochemistry of methanogenesis. *J. Biol. Chem.* **263:**7913–7916.

Tanner, R. S., and R. S. Wolfe. 1988. Nutritional requirements of *Methanomicrobium mobile*. *Appl. Environ. Microbiol.* **54:**625–628.

Bobik, T. A., and R. S. Wolfe. 1988. Physiological importance of the heterodisulfide of coenzyme M and 7-mercaptoheptanoylthreonine phosphate in the reduction of carbon dioxide to methane in *Methanobacterium*. *Proc. Natl. Acad. Sci. USA* **85:**60–63.

Noll, K. M., M. I. Donnelly, and R. S. Wolfe. 1987. Biochemical aspects of methane formation in *Methanobacterium thermoautotrophicum*. *Antonie van Leeuwenhoek* **53:**15–21.

Frankel, R. B., R. P. Blakemore, and R. S. Wolfe. 1979. Magnetite in freshwater magnetotactic bacteria. *Science* **203:**1355–1356.

Romesser, J. A., R. S. Wolfe, F. Mayer, E. Spiess, and A. Walther-Mauruschat. 1979. *Methanogenium*, a new genus of marine methanogenic bacteria, and characterization of *Methanogenium cariaci* sp. nov. and *Methanogenium marisnigri* sp. Nov. *Arch. Microbiol.* **121:**147–153.

Blakemore, R. P., D. Maratea, and R. S. Wolfe. 1979. Isolation and pure culture of a freshwater magnetic spirillum in chemically defined medium. *J. Bacteriol.* **140:**720–729.

The Siren Call of the Sea

I grew up in Massachusetts, a stone's throw from the ocean and a block from the lighthouse in Beverly Harbor in the village of Beverly Cove. The ocean has always held a mysterious attraction for me. I enjoyed long walks along the beaches from Beverly Cove to Manchester, Rockport, and Gloucester in the days when the beaches were open to all and not partitioned off as "private property." The lure of the sea was strong. Having come from a family of modest means, the offer from Purdue University of a full scholarship and residence on campus was too good to turn down—even though it meant temporarily leaving the sea behind. The result was an extraordinarily good grounding in science at an institution where undergraduates truly mattered.

At the time, medicine seemed like the path to choose, and I applied and was admitted to several medical schools. However, a fateful meeting late in my senior year with a graduate student in physical chemistry, who was to become my husband, resulted in an additional year at Purdue University, studying classical genetics instead. After leaving Purdue, my husband and I adventurously attended the University of Washington in Seattle, where we studied together for our doctorates. The work at the University of Washington was a dual track, with all of the coursework in microbiology being taken through the Department of Microbiology in the School of Medicine. But the chance meeting with an extraordinary individual, at the time a newly hired young professor from Scotland—John Liston—led to thesis work in marine microbiology. The field was new. In fact, it was a raw, unfinished science with many paths to follow. My work focused on bacteria associated with marine animals—specifically invertebrates, including shellfish, both mollusks and crustaceans. One of the studies compared microorganisms associated with marine animals from the Rongelap and Eniwetok atolls after the atomic bomb tests. It was a fascinating study because it demonstrated concentration of radioactive elements by microorganisms, work that was confirmed by other investigators in later years and that has relevance for bioremediation in today's society—that is, microorganisms can be used to concentrate and remove radioactive elements from radioactive wastes.

The work at the University of Washington was exciting and clearly pioneering. At that time, women were not welcome onboard ship for oceanographic and fisheries work, especially not overnight. I went on several cruises, but these were always one-day cruises. One of the most exciting trips was a fishing expedition using experimental nets to catch salmon. I had never seen salmon of the size we collected that day. Each of us was allowed to take home a "trophy," which in my case was a sixteen-pound salmon that grilled beautifully in the fireplace of our wee apartment in the University of Washington housing complex.

My interest in marine microbiology expanded to a curiosity about the genetics of marine microorganisms. Very little was known about microbial genetics at the time, and of course in the ensuing two decades, an incredible explosion of information has occurred. The initial work demonstrated the presence of plasmids in marine bacteria, especially those bacteria found in harbors and coastal areas into which effluent from sewage treatment plants and industry was discharged. My students and I demonstrated the association of plasmids with metal resistance in marine bacteria. Furthermore, we were able to demonstrate transfer of plasmids among marine bacteria, not only between marine bacteria but also between terrestrial bacteria entering estuaries and the naturally occurring estuarine and marine bacteria found therein.

The most exciting aspect of the work focused on the systematics of marine bacteria. We were able to show that of those bacteria able to be cultured, the dominant forms were *Vibrio* species. These bacterial species included causative agents of disease in fish, as well as in humans, the most notorious of which, of course, is *Vibrio cholerae*. In the 1960s and early 1970s, my students and I showed that *Vibrio cholerae* O1 is an aquatic bacterium. Of course, this hypothesis was not accepted because the prevailing dogma was that *Vibrio cholerae* was transmitted from person to person, or perhaps carrier to case. Fortunately, we were able to develop specific antibodies to *V. cholerae* O1—highly absorbed polyclonal antibodies and, subsequently, monoclonal antibodies—that we used to demonstrate the presence of *Vibrio cholerae* O1 in water samples from coastal and estuarine waters from which we could not isolate it in culture.

A major limitation to research in microbial ecology has been the inability to isolate, grow, and culture the vast majority of bacteria that are present in the environment. The occurrence of nonculturable, or "uncultured," bacteria has long been known, because enumeration by direct staining has always yielded larger numbers of bacteria than could be cultured, but the nature of the phenomenon was not determined. We reported that selected human pathogens, such as *Vibrio cholerae*, lost the ability to grow on laboratory media after incubation in oligotrophic ocean water or in seawater microcosms in the laboratory for short periods of time (less than one day to three weeks), although cell numbers, by direct microscopic counts, changed little. The implications of these observations proved far reaching in that these pathogens survive in the environment but may not be detected by standard methods. The results of our studies showed that waterborne pathogens that elude detection in the laboratory can retain their pathogenicity and may be "revived" to culturability by animal passage. We showed that from results of animal studies (i.e., rabbit ileal loop analyses), viable but nonculturable *Vibrio cholerae* can be "revived" to the culturable state. Thus, bacteria may not only survive exposure to the marine environment, previously believed to lead to rapid die-off, but they retain important properties, including potential pathogenicity. Therefore, we proposed a resting cell stage for gram-negative bacteria, analogous to spore formation in some gram-positive bacteria.

Much of the work with nonculturable bacteria requires direct counting methods that assay the total numbers of cells in a sample. The direct viable count (DVC) method has been used to estimate metabolically active bacteria populations. DVCs are generally much higher than counts obtained by plate count on agar media and lower than acridine orange direct counts. We were able to demonstrate a strong correlation between DVCs, heterotrophic uptake of radiolabeled substrates, and metabolic activity by microautoradiography, leading us to conclude that the DVC method provides a reasonable estimate of viable bacterial populations, strongly substantiating the existence of viable but nonculturable microorganisms in the environment. The new methods of polymerase chain reaction and gene probing allow direct detection of viable but nonculturable bacteria. More recently, using radiolabeled sulfur substrates, we have been able to demonstrate metabolic activity—that is, protein incorporation of radiolabeled sulfur-containing amino acids.

My exciting basic research work on marine microorganisms extended to include marine biotechnology. In 1983, biotechnology was taking off in a meteoric way, but no effort was being directed toward what I envisioned as the new field of marine biotechnology. This seemed to me a serious shortcoming. I published a paper in *Science* describing the potential of marine biotechnology. Fifteen years later, marine biotechnology is now internationally recognized and pursued with vigor by many countries, including Japan, Norway, France, Thailand, Taiwan, and other countries of Europe, Asia, and Latin America. In the United States, a Center of Marine Biotechnology was established in Baltimore in early 1985 with construction of a facility on the Inner Harbor of Baltimore—the Columbus Center. I was one of the three originators of the Columbus Center, which houses a major research laboratory focused on marine biotechnology and provided an opportunity for public exhibits to describe the excitement of, and developments in, marine biotechnology.

In my current role as Director of the National Science Foundation (NSF), I have the opportunity to bring many of my diverse interests together. NSF is the fulcrum for science and engineering research and education in the United States. The Foundation invests in and across all fields of science, mathematics, and engineering. One example of an area that NSF is now pursuing is biocomplexity, a multifaceted idea that links biology with other areas of science and engineering. It involves working together across disciplines to understand environmental systems. For generations, scientists have studied parts of our environmental system—individual species and habitats—in isolation. It is now time for a better understanding of how these parts function together as a whole, and that approach makes this an especially exciting time to serve as NSF Director and to help foster the advancement of science and the careers of scientists.

When one begins a career, it is not clear where the chosen path will lead. Only a relatively small percentage of graduate students are fortunate enough to spend their entire career in the area in which they did their doctoral thesis. I was one of the lucky ones. Furthermore, this choice, although unforeseen at the time of choosing, proved exciting and at the cutting edge. The message, or perhaps the moral, of my story is that one must choose the path that is of interest and that allows fulfillment and the reward of intellectual challenge. By making a choice based on one's own heart and mind, it is more likely that a rich, rewarding, and exciting lifetime career will be the result.

RITA R. COLWELL was born in Beverly, Massachusetts. She received her bachelor's degree in 1956 in bacteriology and her master's degree in 1958 in genetics from Purdue University. She received her doctoral degree in marine microbiology in 1961 from the University of Washington. She began her career as an Assistant Professor and subsequently Associate Professor of Biology at Georgetown University. She later became Professor of Microbiology at the University of Maryland, a position she held along with serving as Vice President for Academic Affairs of the University of Maryland and then President of the University of Maryland Biotechnology Institute until 1998, when she was selected by President Clinton to become Director of the National Science Foundation. This is the highest position held by either a woman or a biologist in American science. She has been recognized in an exhibition at the Smithsonian Institution for her achievements. She has served on the National Science Board and has been the President of the American Society for Microbiology (ASM), American Academy of Microbiology, International Union of Microbiological Societies, Sigma Xi National Science Honorary Society, and American Association for the Advancement of Science. She received the Fisher Award in Microbiology in 1985 and the Alice Evans Award from ASM in 1985. She has been awarded the International Institute of Biotechnology Gold Medal in 1990 and the Purkinje Medal for Achievement in the Biological Sciences by the Czechoslovakian Academy of Sciences in 1991. She was a Phi Kappa Phi National Scholar in 1992 and won the Andrew White Medal at Loyola College in 1994. She has received the following honorary degrees: Professor Extraordinario from the Universidad Catolica de Valparaiso, Chile, in 1976; Honorary Professor from the University of Queensland, Australia, in 1988; D.Sc. from Heriot-Watt University in Edinburgh, Scotland (1987); Hood College in Frederick, MD (1991); Purdue University in West Lafayette, ID (1993); the University of Surrey, Guilford, England (1995); the University of Bergen, Norway (1999); East Carolina University, in Greenville, NC (1999); the University of Maryland, Baltimore County

(1999); and St. Mary's College, St. Mary's City, MD (1999). Her research focuses on biotechnology, microbial biodiversity, marine and estuarine microbial ecology, microbial systematics, marine microbiology (ecology, physiology, genetics, and fine structure of marine and estuarine bacteria), temperature and high-pressure effects on marine bacteria, microbial degradation, applications of computers in biology and medicine, and the release of genetically engineered microorganisms. She is the author or editor of sixteen books; produced an award-winning film, *Invisible Seas*, a 28-minute color film on marine microbiology; and is the author or co-author of more than 500 papers and articles and approximately 400 published chapters, reports, and abstracts.

The following papers are representative of her publications:

Chun, J., A. Huq, and R. R. Colwell. 1999. Analysis of 16S-23S rRNA intergenic spacer regions of *Vibrio cholerae* and *Vibrio mimicus*. *Appl. Environ. Microbiol.* **65:**2202–2208.

Colwell, R. R. 1996. Global climate and infectious disease: The cholera paradigm. *Science* **274:**2025–2031.

Pommepuy, M., M. Butin, A. Derrien, M. Gourmelon, R. R. Colwell, and M. Cormier. 1996. Retention of enteropathogenicity by viable but nonculturable *Escherichia coli* exposed to seawater and sunlight. *Appl. Environ. Microbiol.* **62:**4621–4626.

Rahman, I., M. Shahamat, M. A. Chowdhury, and R. R. Colwell. 1996. Potential virulence of viable but nonculturable *Shigella dysenteriae* type 1. *Appl. Environ. Microbiol.* **62:**115–120.

Shiba, T., R. T. Hill, W. L. Straube, and R. R. Colwell. 1995. Decrease in culturability of *Vibrio cholerae* caused by glucose. *Appl. Environ. Microbiol.* **61:**2583–2588.

Colwell, R. R. 1994. Biodiversity and release of genetically engineered organisms: A partnership of value. *Curr. Opin. Biotechnol.* **5:**244–246.

Somerville, C. C., and R. R. Colwell. 1993. Sequence analysis of the beta-*N*-acetylhexosaminidase gene of *Vibrio vulnificus*: evidence for a common evolutionary origin of hexosaminidases. *Proc. Natl. Acad. Sci. USA* **90:**6751–6755.

Byrd, J. J., H. S. Xu, and R. R. Colwell. 1991. Viable but nonculturable bacteria in drinking water. *Appl. Environ. Microbiol.* **57:**875–978.

Leahy, J. G., and R. R. Colwell. 1990. Microbial degradation of hydrocarbons in the environment. *Microbiol. Rev.* **54:**305–315.

Roszak, D. B., and R. R. Colwell. 1987. Survival strategies of bacteria in the natural environment. *Microbiol. Rev.* **51:**365–379.

Grimes, D. J., R. W. Atwell, P. R. Brayton, L. M. Palmer, D. M. Rollins, D. B. Roszak, F. L. Singleton, M. L. Tamplin, and R. R. Colwell. 1986. The fate of enteric pathogenic bacteria in estuarine and marine environments. *Microbiol. Sci.* **3:**324–329.

Colwell, R. R. 1977. Ecological aspects of microbial degradation of petroleum in the marine environment. *CRC Crit. Rev. Microbiol.* **5:**423–445.

Mallory, L. M., B. Austin, and R. R. Colwell. 1977. Numerical taxonomy and ecology of oligotrophic bacteria isolated from the estuarine environment. *Can. J. Microbiol.* **23:**733–750.

Colwell, R. R. 1973. Genetic and phenetic classification of bacteria. *Adv. Appl. Microbiol.* **16:**137–175.

36

ERKO STACKEBRANDT

Systematic Challenges

Although I grew up with the option of using a microscope that my father acquired in the 1930s during his medical studies, I did not in my childhood develop a special interest in microscopy. My preoccupation with this instrument, restricted to the observation of a few preparations of pond water and hay extracts, was probably less than that shown by most children who had access to this magnificent toy. My wish to become an architect rather than an observer of natural phenomena evolved during my school time, but the dream to study architecture in Munich was instantly shattered when I failed to pass the entrance examination. Preferences two and three were sports and biology, respectively. Because the entry examination in sports at the Ludwig-Maximilians University of Munich required a test in skiing (to which, being born and raised in the north German flatlands, I was never subjected), I decided to study biology at the same university. What I did not know at the time, because I did not prepare for such a career, was that I could not have found a better place for studying this discipline.

During the following two years, I began to show an interest in zoology and botany, and I caught up with activities most of my classmates had begun a decade earlier, such as collecting insects and plants. As if driven to make up for lost time, my spare time was dedicated to ferns and insects. Within a few seasons I had a nice collection of more than two thousand specimens of beetles, mainly from Bavaria, Austria, and Southern Tirol. With some delay I discovered that my gene pool must contain a trait inherited from my mother's side, one for the serious occupation of collecting. I considered entomology a career worth continuing, but my future took an unexpected twist when I was confronted with the newly introduced discipline of microbiology at the university and the professor who represented it from the end of the 1960s.

Microbiology in Munich was headed by Otto Kandler, a botanist and microbiologist whose scientific emphasis in microbiology was on the physiology and taxonomy of gram-positive bacteria. The department was the world center for the analysis of the chemical composition of the cell wall, and probably no student with a major in microbiology left the department without at least having close contact to picolinic acid or pyridine solvents used in the one-dimensional descendent separation of HCl-hydrolysated peptidoglycan. The scientific atmosphere provided by Otto Kandler and his collaborators, such as Walter Hammes, Franz Fiedler, and Karl-Heinz Schleifer, was fascinating and impressive for students because they could reconstruct the search for natural entities in bacteriology. Equally important, the names of the supervisors could be found on the many publications that originated from this laboratory. We were raised in the spirit that hard work and serious dedication to the scientific task provide excellent cornerstones for one's career. Indeed, when we noticed that many of our older colleagues were offered excellent positions in academia or research institutes, we acknowledged that the selection of the proper team, by choice or by chance, was a second tremendously important prerequisite for a career.

Unlike most of my fellow students, I did not concentrate on cell walls during my diploma thesis in 1972 but was offered another taxonomic subject—namely, the elucidation of the importance of metabolic end products in the classification of coryneform bacteria. Previous analysis of peptidoglycan already pointed toward the discrepancy between classification based on classical phenotypic properties and grouping according to chemotaxonomy. A logical next step was to investigate the taxonomic potential of other characteristics, such as the qualitative and quantitative formation of ethanol, acetate, and lactic acid and the reconstruction of the glycolytic and the pentose phosphate pathways. The results of this study had no major impact on our understanding of the relationships among the coryneform bacteria. For me, however, it provided an introduction to the world of molecular taxonomy and systematics. This period marked the beginning of my search for the interrelationships between bacterial taxa and the implementation of the polyphasic approach to taxonomy. Chemotaxonomy was an important step in the direction of recognizing groups of naturally related species (my favorite textbook at that time was volume 20, No. 4 of the *International Journal of Systematic Bacteriology*). However, it had already been noted that the chemotaxonomic approach was restricted to clustering bacteria on the basis of common nongenetic, not necessarily homologous properties and therefore failed to reveal a hierarchic structure of the bacteria.

Through my interest in the Gram-positive rod-shaped and aerobic bacteria, in 1973 I was offered the position as a curator for coryneform bacteria in the German Collection of Microorganisms (today DSMZ-German Collection of Microorganisms and Cell Cultures GmbH, one of the world's largest microbiological resource centers). The initial collection was founded in 1969, driven by the interests and needs of industry and academia to work with defined and pure cultures. As the collection was decentralized, with the head office in Göttingen, the few scientists in Munich, each of them covering a limited number of strains, had ample time for applying and developing new taxonomic methods.

In parallel to the elucidation of chemotaxonomic data to assess relationships between bacteria, a few groups, such as those headed by John Johnson, Don Brenner, and Josef De Ley (to name a few), developed techniques to determine the molecular relatedness between closely related bacteria. Different approaches to DNA-DNA hybridization were in use, the results of which agreed that strains of a given species showed high DNA similarities. These techniques were applied to hundreds of strains, and the superiority of this approach for demonstrating genomic coherency or heterogeneity was impressive. However, it became quite obvious that the results of DNA reassociation studies were able to reveal only the most recent evolutionary events, with no option to detect the more ancient species history. Although this restriction did not pose a problem for identification of organisms with known genus affiliation, the scientific curiosity of understanding more of the genealogy of bacteria could not be satisfied.

In retrospect, these years prepared me for the understanding that all the approaches used in taxonomic studies were different jigsaw puzzles using pieces that did not yet fit together. Each method was valuable only in the context of what it was developed to be used for; the concentration on certain aspects of the cell had its own merits and revealed important insights, but the overall view of what a prokaryote represents was missing. Later, when I found the saying by the zoologist Theodor Dobszanki that "nothing makes sense in biology—except in the light of evolution," I knew what I had been looking for but had not been able to express.

My search for the determination of more ancestral relationships came after the introduction of ribosomal RNA (rRNA) sequencing to determine phylogenetic relationships. More powerful than the DNA-DNA reassociation approach, the determination of the similarity between the evolutionarily conservative sequence of rRNA and their coding genes allowed microbiologists to determine relationships at the intrafamily level. This was the method of choice to complement the determination of DNA similarities to finally unravel the phylogeny of prokaryotes. So I thought.

I was probably not standing alone when I noticed in 1977 that I had missed an important 1974 publication by Carl Woese and his coworkers. The intrinsic properties of the 16S rDNA sequences allowed the investigation of close relationships (i.e., at the species level) but also very remote relationships (i.e., at the interkingdom level), and all data could be stored and retrieved in a cumulative database. I immediately stopped my work on DNA-rRNA reassociation, and, financed by the German Research Council, I began a postdoctoral year in January 1978 in the laboratory of Carl Woese at the University of Illinois. Without exaggeration, the twelve-month period in Urbana-Champaign completely changed my attitude toward science. This change was mainly because of Carl Woese and his serious commitment to unraveling the evolution of microorganisms, taking no taxonomic relationship for granted and always doubting the results of a century of bacterial taxonomy. It was also because of the 16S rRNA, part of the ribosome and, as Carl Woese had written on his blackboard, "the greatest machine ever built." Within a few months, the broad outline of the phylogenetic structure of most of the main phyla of the domain bacteria was unraveled. After my return to Germany, I continued to work with rRNA, first at the Max Planck Institute for Biochemistry and then at the Technical University in Munich. I continued my career at the Universities of Kiel, the University of Brisbane, and the DSMZ, the German Collection of Microorganisms and Cell Cultures.

Without my background in classical microbial taxonomy, I would probably have converted to a sequence-only taxonomist. In those days this conversion was tempting because of the dominating influence of 16S rRNA sequence analysis in taxonomy. With incredible speed, facilitated by changes in methods, such as the replacement of rRNA cataloguing by the reverse transcriptase method and later by polymerase chain reaction (PCR)–mediated rDNA sequence analysis, the phylogenetic tree grew to encompass any strain for which rRNA/rDNA could be recovered: free-living strains, symbionts, parasites, extremophiles, organelles, eukaryotic microorganisms, and higher evolved forms. Almost all phenotypically defined prokaryotic taxa above the genus level and many intrageneric relationships were demonstrated to be incorrect from a phylogenetic point of view, though previous systematists, who developed their ideas on the basis of the results provided by the methodologies to which they had access during their time, should not be blamed for misconceptions of phylogenetic relatedness.

Overwhelmed by the enormous success story of rRNA sequence analysis, the results of which were later supported by sequence analysis of different evolutionarily conserved molecules and by the ability to work with a hierarchic construct that "made sense in the light of evolution," each taxonomist had to make a choice about how to incorporate the molecular data into his or her view of systematics. There were three categories. The first group hesitated to accept molecular sequencing at all; because the outcomes of these studies were so revolutionarily different from the conclusion shown in *Bergey's Manual of Determinative Bacteriology*, they doubted the validity of the 16S rDNA studies (this group of taxonomists does not exist anymore). The second group favored the idea to define all taxa by small ranges of cut-off similarity values because they put faith in the objectivity of the molecular data. Myself, among many others, belonged to the third group of taxonomists, who agreed to the concept that a species as defined in bacteriology is an artificial construct. Its delineation should be based on a combination of molecular data and phenotypic properties.

The rRNA molecule is too conservative to reflect genomic heterogeneity at the strain level, and it may be too conservative to answer all questions of the ancestry of life. The rate of evolution between the genotype (as measured by the 16S rDNA) and the phenotype (representing many different genes) does not run isochronically, and clusters defined by high 16S rDNA similarity values may show significant differences in phenotype, which does not necessarily correlate with results of DNA reassociation studies.

Granted, in the attempt to define a taxon, the objective data, such as sequences, chemical composition of lipids, peptidoglycan, isoprenoid quinones and the like—end products of carbohydrate fermentation and other phenotypic properties—are treated subjectively. The more complete the data set, the more likely it is to select those characteristics that match the phylogenetic branching pattern most closely. On the other hand, selected phenotypic data, the genes of which cover a broader spectrum of the genome, may help to point out failures of the mathematical algorithms used to phylogenetically relate strains. The polyphasic approach to taxonomy, first outlined by Rita Colwell, has proved successful, and the scientific community has now accepted this laborious way of circumscribing a taxon of any rank. For higher ranks, especially at the levels of families, orders, classes, kingdoms, and domains, the sequence composition and the presence of signature nucleotides provide most important delineation criteria because phenotypic properties shared by all members of a given taxon are rare.

The introduction of molecular sequencing into systematics and the consequent stability of prokaryotic taxa raised the interest of scientists from different bacteriological disciplines. Biochemists, physiologists, morphologists, ecologists, medical microbiologists, and geochemists have discovered the importance of microbial phylogeny and systematics. This interest has led to vigorous interdisciplinary collaborations, and scientists working on different groups of bacteria and from different aspects of the disciplines they represented have been brought together because of the unexpected close relationships between their organisms. Microbial taxonomy can no longer be described as the "ugly duckling" of microbiology, covered with an air of dullness. Within a decade, taxonomy developed into a highly exciting discipline.

The sudden interest in taxonomy and phylogeny can be explained also by the tremendous influence the analysis of rDNA sequences has had in two other disciplines of microbiology: medical microbiology and microbial ecology. Derived and catalyzed by the development of PCR technology, diagnostic methods were developed that began to change bacterial identification in natural samples. Oligonucleotide probing and PCR assays have raised the interest of pharmaceutical companies, some of whom my group has been associated with since 1986 by developing rDNA probes and rapid rDNA-based screening methods.

DNA-based techniques are now complementing immunologic tests in pathology and are widely used to explore the identity and phylogenetic diversity of natural samples. Microbiologists know that the two domains of prokaryotes are grossly underestimated by the description of about four thousand species only. The increasing spectrum of sophisticated monitoring and detection methods will facilitate cultivation of organisms that, until today, resisted being cultured under laboratory conditions. However, even if we restrict our fantasy to a hundred-fold larger species number and we consider that future species description will be more molecular driven and hence less time consuming, the small number of today's systematists still will not be able to seriously handle the expected enormous avalanche of novel strains. The present number and support strategy of microbiological resource centers must be improved to cope with the increased deposition of strains, their continuous quality assessment and authentication, and the implementation of the articles of the Convention on Biological Diversity.

My major scientific driving force originates in the field of tension that exists in the triangle of culture collections (microbiological resource centers), artificial species definition, and the ecological importance of strains. In other words, the question of the "true" nature of a bacterial species, its many possible genetic definitions (depending on the extent of panmixis, the exchange of genetic material), the role of the environment in determining speciation processes, and the transformation of this information into a species concept will keep me from falling into the "resting stage" of a scientist.

Systematists of tomorrow will be neither "molecular only" nor "phenotype only" scientists; they will need to be united and integrated by biological informatics. The last twenty years have witnessed radical changes in our understanding about the role and importance of prokaryotic organisms. Systematists have benefited from the development of molecular methods. There is excitement in being a taxonomist. We are making important contributions to understanding the true nature of the biology of prokaryotes.

ERKO STACKEBRANDT was born in Germany in 1944. He attended the Ludwig-Maximilians-University, where he received his master's degree in biology in 1971 and his Habilitation for Microbiology in 1983. He went on to join the faculty of Christian-Albrechts-University in Kiel, Germany, and subsequently served on the faculty at the University of Queensland in Australia. He is currently the managing director of the German Collection of Microorganisms and Cell Cultures and is also Professor of Microbiology at the Technical University, Braunschweig, Germany.

The papers following are representative of his publications:

Rheims, H., and E. Stackebrandt. 1999. Application of nested PCR for the detection of as yet uncultured organisms of the class Actinobacteria in environmental samples. *Molec. Ecol.* **1:**137–143.

Stackebrandt, E., F. A. Rainey, and N. L. Ward-Rainey. 1997. Proposal for a new hierarchic classification system, *Actinobacteria* classis nov. *Int. J. Syst. Bacteriol.* **47:**479–491.

Jacobi, C. A., B. Assmus, H. Reichenbach, and E. Stackebrandt. 1997. Molecular evidence for association between the *Sphingobacterium*-like organism "*Candidatus comitans*" and the myxobacterium *Chondromyces crocatus*. *Appl. Environ. Microbiol.* **63:**719–723.

Rainey, F. A., N. L. Ward, H. W. Morgan, R. Toalster, and E. Stackebrandt. 1993. Phylogenetic analysis of anaerobic thermophilic bacteria: Aid for their reclassification. *J. Bacteriol.* **175:**4772–4779.

Liesack, W., and E. Stackebrandt. 1992. Occurrence of novel groups of the domain *Bacteria* as revealed by analysis of genetic material isolated from an Australian terrestrial environment. *J. Bacteriol.* **174:**5072–5078.

Liesack, W., H. Weyland, and E. Stackebrandt. 1991. Potential risks of gene amplification by PCR as determined by 16S rRNA analysis of a mixed-culture of obligately barophilic bacteria. *Microb. Ecol.* **21:**188–201.

Stackebrandt, E., R. G. E. Murray, and H. G. Trüper. 1988. *Proteobacteria* classis nov., a name for the phylogenetic taxon including the "purple bacteria and their relatives." *Int. J. Syst. Bacteriol.* **38:**321–325.

Niebel, H., M. Dorsch, and E. Stackebrandt. 1987. Cloning and expression of rDNA from *P. mirabilis* in *E. coli*. *J. Gen. Microbiol.* **133:**2401–2409.

Seewaldt, E., and E. Stackebrandt. 1982. Partial sequence of 16S rRNA and the phylogeny of *Prochloron*. *Nature* **295:**618–620.

Section Ten

PASSING THE TORCH

Louis Pasteur had the wisdom to recognize that "[s]cience, in obeying the law of humanity, will always labor to enlarge the frontiers of knowledge." But, for science to be of value, knowledge must be communicated. What good would van Leeuwenhoek's hours of observing microbes have been if he had not sent letter after letter to the Royal Society? What good would all the discoveries of the last three centuries of microbe hunters have been if they were not transferred to the next generation of scientists? Surely, Pasteur and Koch understood this, for they surrounded themselves with students whom they trained to carry on their works of discovery. They served as teachers and mentors, infecting future microbiologists with the enthusiasm of inquiry and the tools for following in their footsteps of discovery.

In Homer's *Odyssey*, Mentor was Telemachus' surrogate father and counselor during Odysseus' absence. Mentor guided, educated, and protected Telemachus; introduced him to other leaders; and prepared him to assume his adult responsibilities. When the goddess Athena wanted to advise Telemachus, she took the form of Mentor, thereby imbuing Mentor with godlike qualities. It is, perhaps, an overstatement to conclude that mentors must be imbued with godlike qualities.

They must serve as guides to teach, train, and shape careers. In the role of a teacher, they must convey knowledge and share a wealth of career experience. Like the master of an apprentice, they must serve as a role model—correcting mistakes when needed. Like a parent they must provide encouragement and support and know when to release a protégé to carry forth independently. It is not, however, an exaggeration to state that the role of the mentor is among the most important tasks for training the next generation of microbiologists. For as Pasteur declared, "Chance favors the prepared mind."

Moments in the Sun

There have been times in my life when microbiology was not a high priority, but not too many! Microbiology as a career has provided me with much satisfaction and fulfillment. As a dynamic discipline, it has served me as a tool for enticing students into the academic life and illustrating to anyone who would listen the remarkable beauty and diversity of life (which is mainly microbial) on our planet. I am sometimes asked to reflect on the chosen path of my life and am queried about how I would live my life differently should the opportunity arise. Without reservation, I would do as I did—mainly because I really cannot do so but also because I am satisfied with the direction my life has taken.

Career goals were not a high focal point for me when I was growing up in Havana, Cuba. It was assumed, even by me, that I was to follow in my father's footsteps and become a physician. I attended a parochial school whose primary mission was to make sure I did just that. Then, in 1959, all these plans were put on hold when Fidel Castro took control of the Cuban government. In 1962, at sixteen years of age, I immigrated alone to Miami, Florida, in hopes of continuing my education. While at a refugee camp in the Everglades of Florida, I continued my high school education and learning the English language. One of my least favorite subjects was biology, so I started to think that becoming a physician wasn't such a good idea.

In May 1962 I was relocated to Spokane, Washington, where my life as an "American" really began. I enrolled at Gonzaga University as a chemical engineering major, but a combination of homesickness, inability to understand English, the Cuban missile crisis, and the Vietnam war prevented me from focusing on my studies, and as a consequence I did not do too well. Thus, up to this point, my academic career was not the resounding success my parents (and I) expected. From 1963 until 1968, I worked as a physical therapy and surgical orderly in a local hospital. After much deliberation, I decided to pursue a career as a physical therapist.

In 1968, after a brief stint in the Army, I married my wife, Pat (we have now been married 31 years; how time flies when you're having fun!), and returned to college, this time for good, as it turned out, ready to become a physical therapist. At Eastern Washington University, I took the required chemistry and math courses and did well, but my nemesis—biology—continued to hound me. I was frustrated with my inability to dominate this subject, especially genetics, until I took a course in microbiology from a first-year faculty member named Norman Vigfusson (my idol—I call him Dr. Vigfusson even now because I can't enunciate the name *Norman* in his presence). Dr. Vigfusson was a high school teacher who decided to pursue a microbial genetics career well into his thirties (a veritable old man).

My first college-level microbiology course was a career maker. After so much frustration and failures, I fell in love with microbiology. I was absolutely certain that I wanted to be a microbiologist when I grew up. So, I informed my wife, apologized to my parents, and immersed myself into becoming a competent microbiologist. I'm still working at it!

My first exploration into microbiological life was with *Neurospora crassa*. My mentor, Dr. Vigfusson, studied the genetics of the life cycle of this fungus as part of his doctoral dissertation. My senior project was to study the first step in fertilization of the female gamete and to see if there were any chemical signals (pheromones) involved in making the female organs receptive to the male gamete.

To this end, I grew one mating type ("a") on solid culture media, and I harvested the conidia that were to serve as the "male" gamete for fertilization. The other mating type ("A") was grown in a broth culture. After the cultures on broth were mature, they were fertilized and maintained in culture until the sexual organs (perithecia) were present. Afterward, the culture medium was harvested by centrifugation and filter sterilized. This sterile culture filtrate was tested for sterility, concentrated by dialysis, and later used to induce perithecia in single mating–type cultures. We were successful in this endeavor, but not reproducibly so. Thus, my first experience in *avant garde* biology. Instead of frustrating me, however, it made me more determined to find out why. I never did find out why, but I learned a bunch of biology and discipline along the way. I continued my explorations of the sexual development cycle in *Neurospora crassa* for my master's thesis. My efforts culminated in my first publication in *Nature*.

From that point on, I was hooked! After much deliberation and soul searching, in 1972 I entered the laboratory of my second "idol," John J. Taylor at the University of Montana. He, too, to this day, causes me to have speech problems with the name *John* in his presence. To this man, too, I owe a big debt of gratitude. He taught me the scientific method and the precision of scientific nomenclature, but more important, he instilled in me a deep respect for life and for the awesome diversity of microbial life on our planet. This respect and appreciation have motivated me to explore hidden microbial habitats and to study the diversity of microbial life throughout my career.

Under the tutelage of John Taylor, I learned much medical and general mycology. Together we developed a model to study airborne fungal infections using mice as the experimental host and *Emmonsia parva* as the infectious agent. We studied the host's cellular and immune response to inhaled conidia and the development of conidia into spherules (named *adiaspores*, coined to indicate that these structures showed no evidence of cell division) of *Emmonsia parva* in host lung tissue. We published together two papers and four abstracts describing the immune response of mice to inhaled conidia. Along the way we noticed that the inhaled conidia elicited a potent cell-mediated immune response and that normal, healthy mice were virtually immune to disease because their immune response was definitely fungicidal. In immunocompromised mice, however, infections became disseminated and a proportion of those infected succumbed to the infection. We also noticed that the adiaspores showed definite signs of reproduction as they developed buds. In fact, the adiaspores looked much like the tissue phase of *Paracoccidioides brasiliensis* and its multiply budded spores, which resemble a ship's wheel.

I received my Ph.D. in 1974 and was fortunate enough to be offered a tenure-track position in the Department of Biology at California Polytechnic State University (Cal Poly) in San Luis Obispo, which I accepted without hesitation. Through my early years at Cal Poly, I focused on becoming a microbiologist and developing my teaching skills. I moved through the ranks and became a full professor in 1983. Looking back, I do not recall anything remarkable about my career during those years. I guess I was maturing as a microbiologist.

One summer day in 1984, while I was mowing my lawn, I had a thought that, in retrospect, changed my life dramatically. I decided to take a sabbatical leave and rediscover my roots. In 1985 my family and I went to Seville, Spain, where I was to spend my sabbatical working with José Carlos Palomares and Evelio Perea at the Department of Microbiology in the School of Medicine at the University of Seville. Under the tutelage of these two fine scientists and, more important, with their friendship, I learned the rudiment of molecular and clinical microbiology. I actually extracted DNA! Moreover, for the first time in my career, I directed the research of a Ph.D. candidate. During that year, I truly learned the value of a teaching experience that judiciously integrated research in the learning experience. This experience was so powerful and rejuvenating that I resolved to return each year—a commitment I have never violated.

Needless to say, I was heartbroken when I had to return to my professorial duties at Cal Poly, but I was resolved to change my approach to teaching and to include research in the learning experience of my students. I slowly built a molecular biology lab and began to use the senior project graduation requirement at the Biological Sciences Department to teach microbiology to our majors. This laboratory slowly developed into a first-class molecular biology laboratory and has served as a "home away from home" to more than 500 students.

I organized the laboratory so that it included a peer support structure. The laboratory director, in charge of the laboratory and its daily operation, directed the research of graduate students, who in turn supported and directed two senior project students, who in turn served as resources to freshman, sophomore, and junior students entering the lab as research assistants. A central goal of the research projects was to provide undergraduate students with extensive experience in all aspects of a meaningful research project, to pique their scientific curiosity, and to teach them the scientific method.

One such project was again to change my life. It started innocently enough with a simple question. How old can DNA be and still be analyzable? To that end, Hendrik Poinar, a biochemistry undergraduate student at the time, obtained some amber samples from his father (George Poinar), and together we developed a protocol that, in our perception, would prevent the contamination of the amber samples with modern DNA and microorganisms in the environment. At first we were unsuccessful, but after a few tries, we obtained sufficient DNA from bee tissue in amber from the Dominican Republic (20 to 35 million years old) that could be analyzed by DNA hybridization and amplified by the polymerase chain reaction (PCR). We published this study in an obscure journal, but it gave us the hope that a better, more controlled experiment could be designed and carried out under even more stringent conditions.

Hendrik obtained from George Poinar a piece of Middle Eastern amber (from about 120 million years ago) containing a weevil and proceeded to extract its DNA. Fortunately, he was able to extract some DNA from the inclusion, amplify it by PCR, and obtain the nucleic acid sequence of approximately 350 base pairs of the amplicon. This allowed us to carry out phylogenetic studies and determine that the DNA extracted was most closely related to modern weevils. This discovery captivated the nation (and the world) and initiated a great deal of scientific dialog.

The publication of our results in *Nature* coincided with the release of the movie *Jurassic Park*, and it created a great deal of media frenzy. This notoriety was definitely a double-edged sword. It created havoc with the everyday operation of the lab, but the publicity also attracted top-notch students to the laboratory.

There was also controversy, and well there should have been. After all, how can we be sure that the DNA we extracted was indeed from the ancient weevil and not from modern sources? This question has not yet been answered to the satisfaction of the scientific community, and in my estimation, it will never be. How can you use negative results to support a hypothesis?

The issue of laboratory contamination, although troublesome from the scientific point of view, was a motivating force for students to better appreciate the aseptic techniques they had learned in introductory microbiology courses and to apply themselves in designing better experiments, relying on controls to help them interpret the results. So a negative aspect of the research was used to reinforce the learning experience through a well-thought-out and controlled experiment.

They say that everyone has his or her 15 minutes of fame. Well, I have had *two* periods of 15 minutes of fame in my career (and I think two is quite enough!). While Hendrik was working on DNA, another student, Monica Borucki, and I were working on the possibility that viable spore-forming organisms were trapped inside the amber matrix. This possibility insinuated itself in my consciousness after viewing some electron photomicrographs of abdominal tissue from an amber-entombed bee. There, in plain view, were endospores with obvious exposporia and a seemingly intact matrix. They really did not look much different from electron photomicrographs of modern *Bacillus megaterium* endospores.

With this observation as the motivating force, we again toiled over the proper experimental design to obviate environmental contamination of the amber samples. To further reduce the possibility of contamination, we extracted and cultured gut tissue under stringent containment conditions, having a fair idea of the gut's microbial flora in modern bees. From this experiment we isolated a culture of *Bacillus sphaericus* that, to this date and after numerous verifying experiments, we think originated from the amber inclusion. The hypothesis that this organism resided in viable form in the gut of the bee has never been disproved.

After the onslaught of publicity and worldwide attention (and scrutiny) after the publication of our discovery in *Science*, there have been, as expected, a considerable number of challenges to our claims, but in this case, the scientific method has smiled on us. There have been at least three independent verifications of the isolation of a living microorganism from amber. These reports have been shared with senior project and other undergraduate students in my laboratory and scrutinized for the integrity of the experimental design and the results obtained.

Since 1986, when I returned from Spain as a changed man, I have focused on becoming a microbiology teacher and sharing with my students the wonders of microbial life. During that time, I have developed a philosophy of teaching microbiology with the premise that in order to be a successful teacher, you have to be a successful researcher. The scholarships of discovery, application, and teaching have to merge into one to be truly useful. In the classroom, students can learn the fundamentals, but they *must* apply their knowledge, not only in experiments with known outcomes but in a discovery environment. There they have to apply their theoretical knowledge to formulate a hypothesis, develop experiments to test the hypothesis, perform the proper methodology well, and interpret the results. After all, microbiology is a science of "learning by doing."

This philosophy has guided my career since 1986. I have taught and investigated as hard as I could with all the strength that I had. These efforts have been rewarded by the high quality of students who leave my laboratory and their subsequent professional achievements. But my greatest reward has been the recognition of my efforts by my peers, the professional microbiologists. When I was named a Fellow of the American Academy of Microbiology, my research accomplishments were recognized, but when I was awarded the 1997 Carski Foundation Distinguished Teacher Award, I received the greatest accolade of all: I was recognized as a teacher!

RAÚL CANO, Professor of Microbiology in the Biological Sciences Department at California Polytechnic State University in San Luis Obispo, was awarded an associate degree in liberal arts from Spokane Falls Community College, bachelor's and master's degrees in biology from Eastern Washington University in Cheney, and a doctoral degree in microbiology from the University of Montana in Missoula. He is recognized for his outstanding performance in the classroom and for his laboratory training of undergraduates since joining the faculty at Cal Poly in 1974. He has regularly taught a wide variety of undergraduate courses, including introductory microbiology lecture and laboratory classes, medical microbiology, and parasitology lecture and laboratory classes. In addition, he gives a graduate lecture and laboratory in cell biology. He has received several teaching awards from Cal Poly, including the California State University Biotechnology Research Lecturer Award in 1993. In 1994, he was endowed with one of Cal Poly's Outstanding Professor Awards and named the California State University Trustees Outstanding Professor.

The following papers are representative of his publications:

Lambert, L. H., T. Cox, K. Mitchell, R. A. Rossello-Mora, C. Del Cueto, D. E. Dodge, P. Orkand, and R. J. Cano. 1998. *Staphylococcus succinus* sp. nov., isolated from Dominican amber. *Internatl. J. Syst. Bacteriol.* **48:**511–518.

Cano, R. J. 1996. Analysing ancient DNA. *Endeavour* **20:**162–167.

Cano, R. J., and M. K. Borucki. 1995. Revival and identification of bacterial spores in 25- to 40-million-year-old Dominican amber. *Science* **268:**1060–1064.

Cano, R. J., M. K. Borucki, M. Higby-Schweitzer, H. N. Poinar, G. O. Poinar Jr., and K. J. Pollard. 1994. *Bacillus* DNA in fossil bees: An ancient symbiosis? *Appl. Environ. Microbiol.* **60:**2164–2167.

Cano, R. J., H. N. Poinar, N. J. Pieniazek, A. Acra, and G. O. Poinar, Jr. 1993. Amplification and sequencing of DNA from a 120–135-million-year-old weevil. *Nature* **363:**536–538.

38

AMY CHENG VOLLMER

In Love with My Job

"How can I get a job like yours?" This question is posed to me frequently at meetings by interested graduate students and postdoctoral fellows. I think that there are many reasons people ask this question. First, I really like what I do, and I think that is communicated to those with whom I interact. Second, student interest in research is often spurred by a positive undergraduate experience. Finally, people who enjoy teaching would like a job that combines teaching and research to a greater extent than occurs in research universities.

How did I get to this point? By a rather circuitous route, as occurs in most careers. I certainly did not plan to be a tenured faculty member at a selective liberal arts college on the East Coast of the United States. In looking back, it is clear that I followed my instincts and intuitions, which have served me well. In addition, knowledge and reasoning have also been beneficial, but many decisions I made were based on rather incomplete knowledge and sloppy logic. While I believe that scientists and scientific pursuits should be based on understanding and intellectual analysis, emotion and personality (as well as serendipity) play important, but often discounted, roles.

I was born in Las Vegas, New Mexico, a small town with limited cultural diversity. In fact, I was the first Chinese baby in recorded history to be born there. My parents were both chemists who had come to the United States for graduate study in the late 1940s and early 1950s. When I was born, my father was doing postdoctoral research at New Mexico Highlands University. Because of the Korean War, he couldn't find employment in industry because he was Chinese. My parents had intended to return to China after the completion of their graduate studies. However, the China they left behind soon disappeared, and they remained in this country instead.

I remember going to the laboratory with my father on Saturday mornings and watching crystals form in a beaker and getting to collect the crystals in a filter. I liked the physicality of folding filter paper and stirring solutions and working alongside my father. My mother also worked in a laboratory. Occasionally I would get to go to work with her. She worked with many more instruments, and I remember thinking how smart she was to know which of the many thousands of knobs to turn at just the right minute. Not only did I feel comfortable in a laboratory, but I knew that it was where I wanted to be.

My choice of an undergraduate institution was based on several things: small size, fine science departments, favorable male to female ratio, weather, and geography. Rice University fit all of the criteria, and I was lucky enough to gain early decision acceptance. My choice of major at Rice was emotional as well as intellectual. I was good at chemistry, but I just couldn't imagine majoring in it. After all, my parents were chemists, and that would just be too predictable! I loved biology in high school because of a fabulous teacher, Frank L. Stark. His enthusiasm for the diversity of life and its many facets was inspiring. So I went to Rice planning to major in biology. Shortly after I arrived, however, I realized that there was one part of biology that really intrigued me: the part with chromosomes and genes and molecules. (It never occurred to me that it was the *chemical* part of biology that I liked.) One Saturday morning, I was wandering around the Biology Department corridors and met Kathleen S. Matthews, a young assistant professor. When I asked her about biology as a possible major and told her of my interests, she perked up and said, "I think biochemistry is what you want to study." She was part of the newly founded department at Rice, one of the first to offer a comprehensive undergraduate major in biochemistry. That sealed it. Here she was, a cheerful and smart woman, a principal investigator who had her own lab—my first professional role model! As a junior at Rice, I found my way into Dr. Frederick B. Rudolph's lab, working on an undergraduate thesis project. Fred's lab studied purine biosynthetic enzymes from rat liver and bacteria. During one rat liver enzyme purification, I had a negative encounter with a rat, whose tail I had broken before it was to be sacrificed. Fred took the broken, twitching tail out of my hand, looked at my pale face, and said quietly, "Why don't you work on the bacterial enzyme?" Thus, I was introduced to the wonders of bacteria!

I applied to a number of biochemistry programs for graduate work and decided to attend the University of Illinois at Urbana. The people there were the friendliest when I interviewed, and I liked the variety of projects that were being studied. There was also a woman on the faculty; many of the other departments I visited had only male faculty members, although I noticed that many of the senior technicians and instructors were women. At that time, there was no rotation program for first-year students in the department at Illinois. Instead, first-year students met faculty and heard about their research in a small seminar series. It was from such a series that I learned about bacterial cells (in particular *Bacillus subtilis*) and the amazing things they could sense, make, and do. I was interested in biochemistry and intrigued by the context provided by the bacterial cell and its population dynamics.

Working in Bob Switzer's lab was a fabulous experience for my scientific training and also because Bob served as a fine mentor and role model. In contrast to many other biochemistry students, I choose to do most of my coursework in the Microbiology Department, just across the street. I had the unbelievable luck to take classes from Ralph Wolfe, Jeff Gardner, and John Cronan. Their excitement about microbiology was contagious, and I was definitely "infected" for life! I was in the audience in Urbana when Carl Woese outlined his early ideas about the position of the Archaea in evolution. The fertility of the University of Illinois campus for microbiology was something that I certainly have appreciated more each year as I continue to meet microbiologists who were influenced by their experiences there.

Another important aspect of my graduate training was the teaching requirement. All graduate students in the School of Chemical Sciences (where the Biochemistry Department was housed) had to teach one or two semesters. We served as teaching assistants in laboratory sections or tutorial sections, mostly in the general chemistry curriculum. I was lucky enough to teach in the biochemistry lab, because I had taken several biochemistry courses as an undergraduate (such courses were less common for undergraduate students in the 1970s). I loved teaching. Evaluations from students were encouraging because I was able to communicate my knowledge and enthusiasm as well as my technical expertise effectively. I was even recognized with an award by the School of Chemical Sciences; these awards usually went to general chemistry teaching assistants (who taught freshmen) rather than biochemistry teaching assistants (who taught upperclassmen and graduate students—a tougher audience).

After a productive two-year foray into an immunology postdoctoral fellowship, I decided that I preferred the world of prokaryotes, as well as the scientists who studied them. I then began a four-year term as an assistant professor in biology at Mills College in Oakland, California, spending my summers in Dale Kaiser's lab at Stanford University. I became acquainted with myxobacteria and the folks in Dale's lab. Although I enjoyed teaching courses in cellular and molecular biology and immunology at Mills, the environment there was not conducive to a productive research program.

A new opportunity presented itself when our family relocated to the mid-Atlantic area. The DuPont Company had recruited my husband, and there was an opening for a one-year leave replacement for a microbiologist at Swarthmore College. Although I had never taught microbiology, I knew that I would love it, and my research interests were much more closely aligned with the teaching. Because of a number of unforeseen and fortuitous circumstances, I was eventually hired to fill a tenure-track position as a microbiologist. I have now been at Swarthmore for 11 years! I enjoy teaching in the general biology sequence as well as my microbiology and biotechnology courses.

On my sabbaticals from Swarthmore, I have been fortunate to work in the laboratory of Bob LaRossa. His microbial genetics group at DuPont in Wilmington, Delaware, focuses on ways to use basic microbial genetics to benefit the agricultural and pharmaceutical research interests at DuPont. From Bob and his coworkers, I have learned the true value of team work and collegiality. I can also tell my students about the differences between academic and industrial research from a unique perspective.

Being a member of Bob's group has led to many papers and presentations. The research program in my own lab has focused on bacterial stress response, and we have been studying the role of the universal stress protein (UspA), which we believe, based on recent work, acts as a set of "brakes" on most of the stress responses in the cell. Absence of this protein results in an "over-reaction" to stress and drains *Escherichia coli* of its precious energy reserve. Presently, we are investigating the roles of the two genes that show homology to UspA to determine if they also serve in similar capacities. We are also investigating bacterial responses to ultrasound in *E. coli* as part of a study to determine how ultrasound can be best used to sterilize water and contaminated objects. This project is a collaboration with an engineering colleague on campus. The stress response field is exploding with new ideas and discoveries in the areas of microbial ecology, pathogenesis, genetics, physiology, and general microbiology. In 2000, I look forward to serving my colleagues as vice-chair of the fourth Gordon Conference on Microbial Stress Response.

Best of all, at Swarthmore, I am able to combine my research interests with my joy of teaching—of interacting with students in a research laboratory setting. I have been able to engage a number of undergraduate students (as well as high school students) in research. In the forty-three papers and abstracts that I have listed on my curriculum vitae, sixteen have at least one student as a coauthor/presenter. There is nothing more satisfying in my work than contributing to the development of a young investigator. Training includes design and execution of experiments, data analysis, model building, oral and written presentations, and experiencing a general meeting of the American Society for Microbiology (ASM), as well as smaller regional meetings of microbiologists. Finally, our laboratory is a social place. We work hard and we play hard. We have regular group meetings at which every person presents 10 to 15 minutes of material, and, in the tradition of the Switzer lab, we take (and post) pictures of lab members. Anyone who wants to be a scientist because he or she is not a "people person" should probably seek another line of work. I can't imagine an area that requires more effective social interactions than the lab!

As an academic microbiologist, I tell people that I have the best job in the world! I work with fine students in a vibrant department at a fabulous institution. In my field of research (bacterial stress response), I meet interesting and dynamic people who work on ecological, medical, and evolutionary problems. My professional society (the ASM) represents a diverse community of people and works to further the professional development of its members, as well as to encourage potential new members. I learn something every day from colleagues and students, and my enthusiasm for my subject continues to be fueled by these daily interactions. It is hard work to keep up and try to make a contribution. It requires discipline and energy, but I am sustained by numerous supports. I look back on many lucky turns my career has taken and look forward to a microbial world filled with many more lessons!

AMY CHENG VOLLMER was born in 1955 in Las Vegas, New Mexico. She received her bachelor's degree in biochemistry from Rice University in 1977 and her doctoral degree in biochemistry from the University of Illinois in 1983. She was a postdoctoral fellow at Stanford University in the Department of Medicine, Division of Immunology from 1983 to 1985. She was an Assistant Professor of Biology at Mills College from 1985 to 1989. She then moved to Swarthmore College, where she currently is an Associate Professor. She has spent two sabbaticals working at DuPont as a visiting research biologist. Many undergraduate students who have worked in her laboratory have gone on to graduate studies and careers in microbiology and related biological and medical sciences.

The following papers are representative of her publications:

Vollmer, A. C., S. Kwakye, M. Halpern, and E. C. Everbach. 1998. Bacterial stress responses to 1-megahertz pulsed ultrasound in the presence of microbubbles. *Appl. Environ. Microbiol.* **64:**3927–3931.

Vollmer, A. C. 1998. Genotoxic sensors. *Meth. Molec. Biol.* **102:**145–151.

Van Dyk, T. K., D. R. Smulski, T. R. Reed, S. Belkin, A. C. Vollmer, and R. A. LaRossa. 1995. Responses to toxicants of an *Escherichia coli* strain carrying a uspA'::lux genetic fusion and an *E. coli* strain carrying a grpE'::lux fusion are similar. *Appl. Environ. Microbiol.* **61:**4124–4127.

Vollmer, A. C., S. Kwayke, M. Halpern, and E. C. Everbach. 1998. Use of bioluminescent *Escherichia coli* to detect damage due to ultrasound. *Appl. Environ. Microbiol.* **64:**3927–3931.

Vollmer, A. C., S. Belkin, D. R. Smulski, T. K. Van Dyk, and R. A. LaRossa. 1997. Detection of DNA damage by use of *Escherichia coli* carrying *recA'::lux, uvrA'::lux* or *alkA'::lux* reporter plasmids. *Appl. Environ. Microbiol.* **63:**2566–2571.

Belkin, S., T. K. Van Dyk, A. C. Vollmer, D. R. Smulski, T. R. Reed, and R. A. LaRossa. 1996. Monitoring sub-toxic environmental hazards by stress responsive luminous bacteria. *Environ. Tox. Water Qual.* **11:**179–185.

Belkin, S., D. R. Smulski, A. C. Vollmer, T. K. Van Dyk, and R. A. LaRossa. 1996. Oxidative stress detection with *Escherichia coli* harboring a *katG'::lux* fusion. *Appl. Environ. Microbiol.* **62:**2252–2256.

Van Dyk, T. K., D. R. Smulski, T. R. Reed, S. Belkin, A. C. Vollmer, and R. A. LaRossa. 1995. Similar response to adverse environmental conditions by *Escherichia coli* strains carrying *uspA'::lux* or *grpE'::lux* genetic fusions. *Appl. Environ. Microbiol.* **61:**4124–4127.

To Be a Teacher

I can hardly remember a time when I wasn't fascinated with the concept of scientific knowledge and excited by the process of gathering such knowledge. As a kid growing up in rural Pennsylvania, I had little exposure to science in schools. Even in classes labeled *science*, the operating principle was to fill out the work sheet, study the chapter, cram for the test, and then answer the questions with a word order as close to the one in the book as your memory would permit. But even in elementary school, I knew that there was something drastically wrong with this picture.

As a country kid, I spent my summers, my afternoons, even the time walking to and from school, out of doors. The local woods, streams, and farmlands became as familiar to me as my backyard. I noticed even then that I looked at nature differently from other kids. Although most saw things with woody stems and green leaves as "trees," I saw dogwoods, oaks, and maples. I wondered why little grasshoppers looked like adult grasshoppers, but the yellow, white, and black striped larvae of the monarch butterfly resembled the adult not at all. Being curious about such weird stuff made me a bit of an oddball among contemporaries for whom baseball, pinball machines, and what we would now call "just hanging out" seemed to be earth's greatest pleasures.

For my classmates, and my teachers too, there was no distinction between science and the products of science—that is, the factoids that were required to do well on the test. But I suspected that was wrong. First, I discovered that some of their facts were simply not so. The earth, from all that I had read outside of class, really had to be more than a few thousand years old. My ability to pry open chunks of slate from the local coal mine boney pile to reveal lovely tree fern fossils mined hundreds of feet below ground just didn't jibe with a young earth.

Second, the students and the teachers paid no attention to the way the information was obtained, and this, I suspected, was terribly wrong. Medieval monks copied beautiful illuminated herbals generation after generation, without reference to the original plants, so eventually one could no longer recognize the specimen being described. Similarly, teachers perpetuated a mixture of truth, half-truth, and myth without reference to the process by which these "facts" had come to be known. Even in the rare instances in which an "experiment" was performed in class, it was never more than a cookbook iteration of observations made largely by dead white guys, usually hundreds of years earlier.

Even as a kid I knew that you learn about nature by looking at nature. How did I know that oaks and maples were different? By looking at oaks and maples, of course. And how did I know that multicolored caterpillars could construct green chrysalises with golden spikes and emerge later as beautiful monarch butterflies? I held one in captivity and observed the process. Pretty early on, it dawned on me that observation and trial were the surest ways to discover the ways things worked, and it was clear, through failures and unexpected outcomes, that even observation and experiment are not a foolproof path to understanding.

So I survived high school, learning what I could but being bored a lot with the way science was taught. Going off to college, I was sure, would bring a change. Here, for sure, I would get the real scoop on how to do science. To my disappointment, not much changed. The material was higher level and therefore more interesting, but the methods changed hardly at all. It was science by osmosis. Not until my junior year did things substantially change. I elected a class in research parasitology taught by Walter Gallatti at Indiana University of Pennsylvania. Here I had the opportunity to participate in a legitimate research program. The results of our research were presented at a research conference in Syracuse, New York, and for the first time I was surrounded by many students who felt the same way about science that I did. It was an epiphany.

Back at Indiana University of Pennsylvania, I was finishing my class work toward a degree in science education. The irony in that didn't strike me until years later. I was on the road to teach science to the students of the Commonwealth of Pennsylvania, and for all practical purposes, I had only one three-credit class under my belt that had taught me anything at all about the methods of science.

At this point, six months before graduation, my career took an abrupt course correction, and as is often the case, it was because of the intervention of a respected mentor. Willis Bell, my botany teacher and advisor, asked me if I had considered graduate school. I hadn't. With his encouragement I applied to and was accepted by the University of Chicago, his alma mater. The University of Chicago was the first to offer me a fellowship (astonishingly easy to obtain in those halcyon days of science), so in the fall of 1963, I threw my few belongings on a train and headed for Chicago.

The culture shock was extraordinary. The big fish–little pond to minnow-in-an-ocean-of-scholars scenario was almost overwhelming. Fortunately friends and sympathetic professors helped make the transition survivable.

The next five years of my graduate program were some of the most stimulating and challenging of my life. I was fortunate in my choice of mentors and graduate advisor. I probably owe my career as a teacher of microbiology to Edward Garber, who introduced me to the sometimes logical, sometimes serendipitous, often quirky nature of scientific research and to the simple eloquence and nobility of teaching with style and sincerity. To my thesis advisor, Robert Tuveson, I owe a great debt for his enthusiasm, insight, and patience. I had had many skilled and dedicated English instructors, but it was Bob Tuveson who, in accepting nothing less than perfection, finally taught me how to communicate in speech and writing. The list of excellent scholar teachers at Chicago was extraordinarily long, but these two mentors were special because they recognized that not all their student products were destined for a research career. Many, perhaps most, would enter institutions in which their principal responsibility would be teaching. Tuveson and Garber acknowledged and encouraged that choice, affirming the decision while modeling the role.

In light of the job market of recent years, my move to Penn State was remarkably easy. After a shaky adjustment to my first real position, I found myself at the Penn State Altoona College. The Altoona College was and is a branch campus of the extensive nineteen-campus Penn

State system. Altoona is a small city with a history coupled closely to the rise of the railroads in the United States. The lingering death of the Pennsylvania Railroad forced the city to find a new niche and a new identity. The community has not always seemed to be able to decide which course of action is in its own best interests, but in one respect Altoona had a clear vision. In 1936, the city petitioned Penn State University to bring to the city a branch campus. The campus, after a few years in a retired elementary school building, purchased an abandoned amusement park called Ivyside Park and began the process of building a college to meet the needs of the community and the state. In a half century the campus has grown in size and quality and is now a degree-granting college in its own right.

For nearly 30 years, I have been privileged to be a member of the faculty of Penn State in Altoona. I have conducted research in the physiology of mycotoxin production and the incidence of pollution in natural springs in Pennsylvania. The collaborations with colleagues engendered by this research have given me great enjoyment. But by far my greatest pleasure has been in introducing my undergraduate students to research in microbiology through my classes and in independent study. If asked to define the nature of good science teaching, I would suggest that the best method integrates the content, process, and ethics of scientific discourse.

With the coming of each new fall semester, I view my classes in light of what I have learned about my profession in the previous year. The year 1999 was a particularly productive one, and although I confess that none of the following things is particularly unique, for some reason their importance has been impressed on me in a way that demands new attention. These, then, are my goals for the new academic year.

- To make myself more accessible to my students and to make my office a place that students enjoy visiting.
- To give increased emphasis to active and collaborative learning. The process of learning science should model the methods practiced by scientists. A less didactic, more participatory approach is indicated, particularly one involving students working in teams.
- To be a better mentor. I will attempt to serve all my students well while keeping an eye open for the exceptional. Culture these students carefully. They are the stocks from which the next century's scientific breakthroughs will emerge.
- To encourage thinking, not recitation. I will assign problems that

challenge and exercise the minds of my students, not their capacity to memorize.

- To use controversy as a teaching strategy. I must remember that all learning is active; in some sense, one cannot be taught, one can only learn. My function as a teacher is, then, to provide stimulating educational environments that elicit learning on the part of my students. Controversy not only stimulates interest; it elicits argument and debate.
- To remember that my responsibilities as a teacher extend beyond the college classroom door. Precollege students, their teachers, scout troops, clubs, and community organizations also have a legitimate claim on my time. Outreach to these groups is good policy and helps to maintain the educational pipeline, while keeping it filled with high-quality students.

I am convinced of the central position of teaching in the field of microbiology. Teaching *is* scholarship, not peripheral to it. Each of us as a microbiologist owes our profession to one or more mentors who took an interest in us as developing scientists. We owe these mentors a debt. Some of us repay it through our research, others by maintaining the flow of quality students through our teaching. It is one of my greatest pleasures to be visited by former students now in allied health, medicine, industry, or research and to find them excited about their work and to know that I had a part in their success. It is times like this that I understand and begin to believe that, as Stephen Jay Gould maintains, "the noblest word in the English language is *teacher*."

JOHN LENNOX was born in 1941 in Parkersburg, West Virginia. He received his bachelor's degree in biology in 1963 from Indiana University of Pennsylvania, his master's degree in botany in 1965 from the University of Chicago, and his doctoral degree in botany in 1968 from the University of Chicago. He was Assistant Professor of Genetics at Penn State University Park from 1968 to 1971. He then joined the faculty of Penn State in Altoona, where he is now Professor of Microbiology. He was instrumental in making teaching microbiology an integral part of the American Society for Microbiology (ASM) annual meeting and has been chair of the teaching division of ASM. He is developer of a program in critical thinking, microbiology, environmental biology, nature in winter, and dinosaurs for children in the nine-to-thirteen-year age range. He has received several honors in recognition of his teaching, including the National Association of Biology Teacher's Two-Year College Teaching Award in 1995, the Penn State University Continuing Education Recognition Award for academic excellence in 1992, and the Amoco Foundation Award for excellence in teaching performance in 1984.

The following papers are representative of his publications, many of which are aimed at college teaching:

Lennox, J., and M. Duke. 1997. An exercise in biological control. *Amer. Biol. Teacher* **59:**36–43.

Lennox, J. 1991. Extracellular digestion. *Amer. Biol. Teacher* **53:**376–379.

Lennox, J., and T. Blaha. 1991. Leaching of copper ore by *Thiobacillus ferrooxidans. Amer. Biol. Teacher* **53:**361–368.

Lennox, J. E., and M. J. Kuchera. 1986. pH and microbial growth. *Amer. Biol. Teacher* **48:**239–241.

Lennox, J. 1985. Those deceptively simple postulates of professor Robert Koch. *Amer. Biol. Teacher* **47:**216–221.

Lennox, J. 1985. Osmotic pressure, bacterial cell walls, and penicillin: A demonstration. *J. Coll. Sci. Teaching* **14:**106–109.

Lennox, J. E., and L. J. McElroy. 1984. Inhibition of growth and patulin synthesis in *Penicillium expansum* by potassium sorbate and sodium propionate in culture. *Appl. Environ. Microbiol.* **48:**1031–1033.

Lennox, J. E., S. E. Lingenfelter, and D. L. Wance. 1983. Archaebacterial fuel production: Methane from biomass. *Amer. Biol. Teacher* **45:**128–138.

Lennox, J. 1980. *Agrobacterium* and tumor induction: A model system. *Amer. Biol. Teacher* **42:**160–166.

Lanier, W. B., R. W. Tuveson, and J. E. Lennox. 1968. A radiation sensitive mutant of *Aspergillus nidulans. Mut. Res.* **5:**23–31.

Lennox, J. E., and R. W. Tuveson. 1967. The isolation of ultraviolet sensitive mutants from *Aspergillus rugulosus*. *Radiation Res.* **31:**382–388.

Mentoring the New Generation of Scientists

Notwithstanding the larger, less than exemplary dimensions of life growing up as an African-American in a comparatively impoverished farming community in Alabama in the 1940s and 1950s, rural life afforded a rich engagement with nature, and I spent much time pondering the many natural examples of living systems with a particular emphasis on why and how! This rather generic interest found a structured placement in elementary and secondary science and mathematics classes. In an otherwise quite limited educational environment, I was the beneficiary of exceedingly talented and committed teachers—ironically, one of the few benefits of a totally racially segregated public school instructional workforce. Among the many outstanding teachers, I acknowledge the extraordinary efforts of Sterling Wallace, an upper-grade school teacher, and Mr. Sanders and Ms. Hope, my high school chemistry and physics and algebra-geometry teachers, respectively. With little to no financial resources, I initially attended Tuskegee Institute (now Tuskegee University) as a five-year "work-study" student and subsequently transferred to Miles College, a small historically black college in Birmingham. Having elected to major in biology, I had the good fortune to be assigned as a laboratory assistant for most of my undergraduate studies. This multiyear experience, along with excellent professors in organic chemistry, biochemistry, genetics, classical structural studies, and histology, afforded me a productive and enjoyable undergraduate tenure. I was, by this time, quite convinced that I could make the effort to become a scientist!

After being rejected by biological sciences departments at several major research universities, I quite belatedly applied to and was admitted to Atlanta University (now Clark-Atlanta University), which at that time had only a master's degree program. In retrospect, this was a critical development, for I was able to extend the depth and breadth of my knowledge in selected subject areas, conduct original research as an integral part of the degree requirements, serve as a teaching assistant at a nearby college, and, most important, be in the presence of African-American scientists of unusual abilities and self-confidence. Most notable among these faculty members was Lafayette Frederick, who maintained an active research program and was a mentor of the first rank. Because of Frederick's efforts, several other Atlanta University graduate students and I—by design or otherwise—were the first African-American students to present papers at the Southeastern Scientific Society's annual meeting. Although of no particular import to me at the time, that historic transaction on the campus of Emory University in the spring of 1963 represents an unpayable debt to Lafayette Frederick.

Armed with the aforementioned experiences and one year of employment as an instructor in the Biology Department at Atlanta University, I journeyed to Purdue University for doctoral studies. My going to Purdue was only slightly more informed than accidental, for I accepted admission because it was the first institution to offer both admission and a teaching assistantship. With regard to the study of microbiology, it ranks as the best decision made in my career, for it was the same fall of 1964 that Ed Umbarger joined Sewell Champe, Art Aronson, and Fred Neidhardt—to name a few of the faculty members of the Department of Biological Sciences.

I had the distinct pleasure to join the laboratory of Fred Neidhardt as a graduate student, and my research focused on the regulation of amino acid and aminoacyl-tRNA synthetase synthesis in *Escherichia coli*. This was a rare opportunity to learn microbial genetics, biochemistry, and physiology under the supervision of an intellectually demanding, professionally sophisticated, and seasoned mentor. On the completion of my thesis work, I accepted an American Cancer Society postdoctoral fellowship in the Department of Biochemistry at the State University of New York at Stony Brook with Martin Freundlich. As a postdoctoral researcher, I continued studies of gene regulation focusing on the role of aminoacylated cognate transfer RNAs as effectors in isoleucine, valine, and leucine biosynthesis.

In the spring of 1969, I joined the faculty of Atlanta University instead of taking offers made by Purdue and Johns Hopkins. I did so with the hopes of embedding mentorship of African-American graduate students as an integral part of my aggregate professorial activities. In a very short period of time, I was soon mentoring a very large pool of graduate students, not because of anything I brought to the process, I suspect. The students' judgment was that I was the youngest, newest, probably most naive and therefore easiest member of the faculty. Thus, there was an enormous gathering. I was also fortunate in successfully competing for a research grant that established my laboratory.

A year later, Henry Koffler, then head of the Department of Biological Sciences at Purdue, visited me in Atlanta and again extended the offer of an assistant professorship at Purdue. It was novel, because I was to become the first African-American member of not only the department but also the entire School of Science. And I was recruited to replace my own mentor, Fred Neidhardt, and inherit his lab as he left to join the University of Michigan. I entered an unusual set of negotiations with Dr. Koffler. I required that an agreement be made between Atlanta University and Koffler that my lab be transferred to Purdue in total—including students. Purdue agreed to allow all of my graduate students to either transfer to Purdue with me to complete their research *in absentia* or to apply for admission to its department. Thus, some of the students received their degrees from Atlanta University and others—the majority—ultimately received their degrees from Purdue.

Once the transfer took place, Ed Umbarger, with whom I had taken several courses and who served on my Ph.D. committee, assumed the role of mentor to me as a young assistant professor. This unusual relationship, which extended over my ten-year employment (absent one year spent in the Biology Department at MIT), become one of the most rewarding and cherished experiences of my life. That he served as essentially coadvisor to my students in no small measure enabled the department to become one of the leading producers of African-American Ph.D.s in microbiology. It was then quite appropriate that Umbarger should present me to the faculty as a Distinguished Alumnus of the School of Science in 1997.

Mentorship, independent of what Fred Neidhardt had exemplified for me, became an all-consuming process. In fact, it became an intellectual continuum for the distinction between courses, seminars, research, qualifying preliminary examinations—along with a special ongoing discussion labeled "the weekly research meeting." We talked about the research but then spent hours in a process that can only be termed "a class" in which all of the participating students had to tolerate my synthesis of what I thought they should know on a weekly basis that encompassed microbial genetics, biochemistry, and physiology. Reflecting on it, it probably served the students in the sense that it translated discrete bodies of knowledge acquired from something called "courses" into what I would call a continuum of knowledge of science since students were assigned to assist their more junior colleagues. I would hope that role modeling in the case of these African-American students was of some value.

One powerful ongoing event at Purdue is the intergenerational continuum of undergraduate, graduate, and professional activity that is under way and has served in excess of fifty Ph.D. recipients to date and hundreds of undergraduates. I think it's probably fair to say that the efforts of my students created a learning opportunity in that department that has and will be sustained for a long period.

Very early in their graduate education, I introduced the students to what I call "essential components of science professionalism"—that is, among other things, they were expected to regularly make presentations of their research findings within the department and the university and at large local and national meetings—the American Society for Microbiology (ASM), in particular. Beyond this, I think the most important of the activities in which we engaged was the creation of an intellectual ambiance that was nurturing to the students who came with different preparations, different interests, and, I would argue, different perceptions of their self-worth—they were placed in an environment that simply disallowed those differentials. If, in fact, there were genuine deficiencies in physical, mathematical, or chemical areas, one addressed them. The idea was that in graduate school, a successful matriculate depended on healthy self-esteem. Any other notion was inconsistent. To me, effort was the critical factor in determining what was accomplished by a student.

After Purdue, I moved on to administrative positions at several universities, the National Science Foundation, and the National Institutes of Health, the latter being the agency that awarded me a predoctoral fellowship, a postdoctoral fellowship, and a Career Development Award and supported my research for eighteen years. The transition from academia to government service was difficult; it was a move to an alien environment that had different core values.

The quite problematic, perhaps even unlikely, circumstances that attended my early education and budding interest in science do not equal a guide to career development as a microbiologist. Choices aside, I was mostly determined and quite lucky! Nonetheless, beyond the personal dimensions of the intellectual challenges of my career, this journey is heightened in significance by the award of the ASM Hinton Research Training Award (1998), which speaks to the commitments, abilities, and accomplishments of my former graduate students and the many professional colleagues who encouraged and supported my efforts.

Throughout my scientific career, in academia and government, I thought that we had to emphasize the abilities of African-American students rather than their past performance. We need to create an educational environment for a sustained learning experience—a true institutional transformation that allows African-American and other minority students to develop their full intellectual potential and to take their proper places within the scientific community. Mentoring is the key to this effort. Just as Umbarger was my mentor and he and I were our students' mentors, so the others we trained have gone on to train yet more scientists to take on the endless search of discovery. The energy and hard work of students is the key to creating an intellectual ambiance that places emphasis on scientific discovery and finding truths. My own efforts have been amplified by my students who went on to replicate my mentoring model and to help train the next generation of African-American microbiologists. My own contribution has been to convert young people's sense of what is possible by providing a role model.

I have maintained for the thirty-plus years of my professional career that each of us can contribute to the eradication of our society's insulting and infinitely destructive exclusion of African-American and other minority individuals from participation in the scientific enterprise. This was and continues to be our challenge and our opportunity.

LUTHER WILLIAMS has a distinguished record as a scientist, educator, and administrator. He was born in 1940 and reared in Sawyerville, Alabama, a rural community in the southwestern portion of the state. He earned his bachelor's degree in Biology at Miles College in Birmingham in 1961, his master's degree in Biology from Atlanta University in 1963, and his doctoral degree in microbial physiology from Purdue University in 1968. His commitment to science training and research is evident from the positions he has held during the past 22 years. Since the mid-1970s, he served as assistant provost and director of the Minority Center for Graduate Education at Purdue; dean of the graduate school at Washington University; vice president for academic affairs and dean of the graduate school at the University of Colorado at Boulder; president of Atlanta University; and deputy director of the National Institute of General Medical Sciences of the National Institutes of Health. In 1990 he became Assistant Director of the Directorate for Education and Human Resources at the National Science Foundation (NSF); he held that position until 1999. At NSF his responsibilites included science, engineering, and mathematics programs for women, minorities, and persons with disabilities. He currently is a visiting scholar at the Payson Center for International Development and Technology Transfer at Tulane University. During his active professional research career at Purdue, the Massachusetts Institute of Technology, Washington University, and Atlanta University (now Clark Atlanta University), he trained over twenty graduate students, thirteen of whom are African-Americans. He is a fellow of the American Academy of Microbiology, a fellow of the American Academy for the Advancement of Science, and the recipient of an honorary doctorate of science degrees from Purdue University, the University of Louisville, Capitol College, Bowie State University, and Tuskegee University. He received the Presidential Meritorious Rank Award in 1993. In 1998, he was the first recipient of the William A. Hinton Research Training Award for outstanding and significant contributions toward fostering the research training of underrepresented minorities in microbiology.

The following papers are representative of his publications:

Williams, A. L., L. S. Williams. 1985. Control of isoleucine-valine biosynthesis in a valine-resistant mutant of *Escherichia coli* K-12 that simultaneously acquired azaleucine-resistance. *Biochem. Biophys. Res. Comm.* **131:**994–1002.

Davidson, J. P., D. J. Wilson, and L. S. Williams. 1982. Role of a *hisU* gene in the control of stable RNA synthesis in *Salmonella typhimurium*. *J. Molec. Biol.* **157:**237–264.

Davis, L., and L. S. Williams. 1982. Altered regulation of isoleucine-valine biosynthesis in a *hisW* mutant of *Salmonella typhimurium*. *J. Bacteriol.* **151:**860–866.

Fayerman, J. T., M. C. Vann, L. S. Williams, and H. E. Umbarger. 1979. *ilvU*, a locus in *Escherichia coli* affecting the derepression of isoleucyl-tRNA synthetase and the RPC-5 chromatographic profiles of $tRNA^{ile}$ and $tRNA^{val}$. *J. Biolog. Chem.* **254:**9429–9440.

Levinthal, M., M. Levinthal, and L. S. Williams. 1976. The regulation of the *ilvADGE* operon: evidence for positive control by threonine deaminase. *J. Molec. Biol.* **102:**453–465.

Fitzgerald, G., and L. S. Williams. 1975. Modified penicillin enrichment procedure for the selection of bacterial mutants. *J. Bacteriol.* **122:**345–346.

Coleman, W., Jr., E. L. Kline, C. S. Brown, and L. S. Williams. 1975. Regulation of branched-chain aminoacyl-transfer ribonucleic acid synthetases in an *ilvDAC* deletion strain of *Escherichia coli* K-12. *J. Bacteriol.* **121:**785–793.

Jackson, J., L. S. Williams, and H. E. Umbarger. 1974. Regulation of synthesis of the branched-chain amino acids and cognate aminoacyl-transfer ribonucleic acid synthetases of *Escherichia coli*: A common regulatory element. *J. Bacteriol.* **120:**1380–1386.

Coleman, W. G., Jr., and L. S. Williams. 1974. First enzyme of histidine biosynthesis and repression control of histidyl-transfer ribonucleic acid synthetase of *Salmonella typhimurium. J. Bacteriol.* **120:**390–393.

McGinnis, E., A. C. Williams, and L. S. Williams. 1974. Derepression of synthesis of the aminoacyl-transfer ribonucleic acid synthetases for the branched-chain amino acids of *Escherichia coli. J. Bacteriol.* **119:**554–559.

Levinthal, M., L. S. Williams, and H. E. Umbarger. 1973. Role of threonine deaminase in the regulation of isoleucine and valine biosynthesis. *Nature* **246:**65–68.

Archibold, E. R., and L. S. Williams. 1973. Regulation of methionyl-transfer ribonucleic acid synthetase formation in *Escherichia coli* and *Salmonella typhimurium*. *J. Bacteriol.* **114:**1007–1013.

Williams, A. L., D. W. Yem, E. McGinnis, and L. S. Williams. 1973. Control of arginine biosynthesis in *Escherichia coli*: inhibition of arginyl-transfer ribonucleic acid synthetase activity. *J. Bacteriol.* **115:**228–234.

Williams, A. L., and L. S. Williams. 1973. Control of arginine biosynthesis in *Escherichia coli*: characterization of arginyl-transfer ribonucleic acid synthetase mutants. *J. Bacteriol.* **113:**1433–1441.

Williams, L. S. 1973. Control of arginine biosynthesis in *Escherichia coli*: role of arginyl-transfer ribonucleic acid synthetase in repression. *J. Bacteriol.* **113:**1419–1432.

Yem, D. W., and L. S. Williams. 1973. Evidence for the existence of two arginyl-transfer ribonucleic acid synthetase activities in *Escherichia coli. J. Bacteriol.* **113:**891–894.

McGinnis, E., and L. S. Williams. 1972. Regulation of histidyl-transfer ribonucleic acid synthetase formation in a histidyl-transfer ribonucleic acid synthetase mutant of *Salmonella typhimurium*. *J. Bacteriol.* **111:**739–744.

Archibold, E. R., and L. S. Williams. 1972. Regulation of synthesis of methionyl-, prolyl-, and threonyl-transfer ribonucleic acid synthetases of *Escherichia coli. J. Bacteriol.* **109:**1020–1026.

McGinnis, E., and L. S. Williams. 1972. Role of histidine transfer ribonucleic acid in regulation of synthesis of histidyl-transfer ribonucleic acid synthetase of *Salmonella typhimurium*. *J. Bacteriol.* **109:**505–511.

McGinnis, E., and L. S. Williams. 1971. Regulation of synthesis of the aminoacyl-transfer ribonucleic acid synthetases for the branched-chain amino acids of *Escherichia coli*. *J. Bacteriol.* **108:**254–262.

Yem, D. W., and L. S. Williams. 1971. Inhibition of arginyl-transfer ribonucleic acid synthetase activity of *Escherichia coli* by arginine biosynthetic precursors. *J. Bacteriol.* **107:**589–591.

Williams, L. S., and F. C. Neidhardt. 1969. Synthesis and inactivation of aminoacyl-transfer RNA synthetases during growth of *Escherichia coli*. *J. Mol. Biol.* **43:**529–550.

Williams, L. S., and M. Freundlich. 1969. Role of valine transfer RNA in control of RNA synthesis in *Escherichia coli*. *Biochim. Biophys. Acta.* **179:**515–517.

Chrispeels, M. J., R. F. Boyd, L. S. Williams, and F. C. Neidhardt. 1968. Modification of valyl tRNA synthetase by bacteriophage in *Escherichia coli*. *J. Mol. Biol.* **31:**463–475.

FREDERIC K. PFAENDER

Environment and Education

What would be considered the most significant thing to happen to microbiology in the last quarter of the twentieth century if we were to ask that question twenty-five years from now? I posed that question to a friend, who thought the revolution in molecular tools that led to our understanding of microbial diversity would be the number one event. I countered with my thought that the recognition of the contribution of microorganisms and their metabolism to global processes by chemists, geologists, engineers, and even physicists would likely have a more lasting impact. I admit that my view is more than somewhat biased by where I work and what I do. I have spent my entire professional life in the Department of Environmental Sciences and Engineering at the University of North Carolina—twenty-eight years trying to convince the engineers and chemists with whom I work that microbiology is more than pathogens in drinking water. In the 1990s, they seemed to be catching on.

Looking back on my career, I had the advantage of growing up in a middle-class family in Southern California in the 1950s and 1960s. This was the period when Southern California was building its reputation as a "land of fruits and nuts," a culture open to all kinds of ideas and activities. My first strong influence was my German immigrant father. He was a brilliant mechanical engineer who gave up a chief engineer position with a major corporation to start his own business, building machinery for the container industry. His company never got very big because he couldn't give up the pure pleasure of building his designs with his own hands. Just good enough was never good enough for him. Perhaps more significantly I learned that joy in what you are doing is maybe more important than money. There was some disappointment when I didn't follow him down the engineering trail, but my interests were somewhere else.

I remember little about public school other than that I had a really good time. But notably there was Francis St. Lawrence. He was teaching middle school science despite an Ivy League biology Ph.D., because he believed that you had to interest people in science while they were young and that doing so was important. He used all kinds of what we now call "active learning strategies" before anyone had put a name on them.

When I completed high school, I hadn't a clue what I wanted to do with my life. Financial constraints drove me to what was then a major Southern California phenomenon, the two-year city college. I dabbled in the liberal arts for a year but found it unsatisfying. A required chemistry course rekindled an interest in science that had been buried by a series of abysmal high school science teachers. From then on it was clear where I wanted to go—medical school! Armed with a little knowledge and the arrogance of youth, I moved to what was then called California State College at Long Beach to finish premed. I read in the catalog that most premeds majored in microbiology. I didn't know what that was but was confident I could do anything.

In 1964, Cal State Long Beach had one of the largest microbiology departments in the United States, with nearly 400 majors. Most were destined for positions as medical technologists. A great faculty, whose careers were based largely on their teaching skills, populated the department. The curriculum was very broad and required course work in all aspects of the science, from medical, physiology, genetics, parasitology, and serology to food and industrial microbiology. Frank Swatek became my advisor, mentor, and friend. He taught me what being a professor was all about and the joys of dealing with students. Within a year I gave up the silly idea of medical school and formed a new goal for my career. I wanted to be a microbial ecologist, do research on microbial mediated transformations in the environment, and teach at a major university—but not in a microbiology department. Somehow that is how it all turned out.

I wisely chose Dr. Martin Alexander at Cornell University as my next mentor to guide me to a Ph.D. I was at Cornell during the years when all of the environmental sciences were making great strides in both identifying environmental problems and understanding their basis. I found the cutting edge of microbiology, ecology, and biochemistry. My work on DDT degradation provided both the opportunity and necessity of using my microbiology along with analytical chemistry (even some synthetic organic chemistry, which I did not enjoy) to look at metabolism by intact communities of organisms, a bent that has persisted over the years. It was a great time to be in science.

My career took important turns while at Cornell. Marty Alexander showed me how a professional scientist behaves. I didn't and don't always follow the example, but it is a good one to have. He also demonstrated how reading broadly prepares you to see where the new frontiers are likely to be, and if you are prepared you can be one of the leaders into those new areas. Ray Horvath, then a postdoc in our laboratory and one of the most clever experimentalists I have ever encountered, taught me how to think about and design an experiment. Unfortunately, lifelong medical problems cut short what promised to be a brilliant career. It was in Marty Alexander's lab that I met Sheila Schwartz, the funny blonde girl from Minnesota who has changed and enriched my life more than anyone.

At the end of three wonderful years at Cornell came perhaps the major turning points in my career. I had to choose between going back to California to do a postdoc and taking a real job. I chose to take an assistant professor position in the Department of Environmental Sciences and Engineering at the University of North Carolina. I replaced both Robert Mah and James Staley, who left simultaneously. I started to build a research program on biodegradation without the preparation of a postdoctoral experience. This made for a fairly slow start.

As the department's only microbiologist, I found myself involved in a variety of projects, including drinking water safety, virus detection techniques, and pesticide analysis. My research group began developing techniques that used radiolabeled compounds to measure pollutant metabolism in a variety of environmental matrices. Initially, the lure was the ability to measure metabolism at concentrations close to those that actually occur rather than the orders of magnitude higher required by the then current state of the analytical chemistry art. We eventually figured out how to handle the kinetics and to do mass balances. Being able to measure the metabolism of intact communities was also possible. We continue to use much more sophisticated versions of these same techniques today. Our initial work was in surface and estuarine waters with contaminants like pesticides and, everyone's favorite, petroleum.

I have always thought that for microbial transformations to occur in any particular environment, there needed to be a combination of the right organism, the presence of the substrate chemical in a form the microbes could access, and a set of favorable environmental conditions that will dictate whether and how fast degradation may occur. The goal of our research in diverse environments has been to understand how the environmental component regulates what happens. An early manifestation of this goal was the work on community adaptation. Over the years what we learned was that adaptation generally meant changes in the community of organisms, selecting for ones that mineralize the compound. In unadapted communities, the compound of interest may still be metabolized, but intermediate products, rather than CO_2, are more common. Until very recently most researchers—us included—have not been very good at finding these metabolic products in natural ecosystems. The advent of new analytical tools has made this search easier.

I took a detour through academic administration, spending seven years with titles like "director" and "chairman" after my name. Aside from detracting from my research, these years were interesting and taught me a good bit about managing people and organizations. I also spent six of those same years as a member of the Environmental Protection Agency Science Advisory Board's Ecological Processes and Effects Committee as their token microbiologist. My preference for looking at whole systems from several perspectives turned out to be valuable in addressing the difficult issues the agency must face.

Over the last decade our research focus has moved out of the water into the soil environment. Accompanying our group on this migration has been humic acid chemist, best friend, tennis and golf partner, colleague, and neighbor Russell Christman. The triad of organism, chemical, and environment-regulating degradation has not changed, but the setting and the questions have. We are now beginning to understand that the interaction of the chemicals with the soil organic carbon, which is the overwhelming regulator of availability and fate, is strongly influenced by the activities of the soil microbial community. We don't understand the details yet, but we are well on the way. Our microbiology, combined with Russ' knowledge of soil organic materials, offers the potential to understand how these two major components of the soil microenvironment relate to each other. Until we have this understanding it will be very difficult to figure out how the availability of the hydrophobic contaminants is regulated both for toxicity and for developing strategies for clean-up.

Interaction with students has always been the thing about a university I enjoyed the most. In fact, for most of us, particularly those of us lucky enough to deal mostly with graduate students, it is why we are here. I suspect it is why, when asked by my wife if I ever intend to grow up, my response is "Why would I want to do that?" Classroom teaching has always been fun for me. I can honestly say I have enjoyed the association with every one of my graduate students.

One of the few skills I claim as valuable is the ability to do multiple things simultaneously. There are some who claim that this is an inability to focus, while others just suggest attention deficit disorder. Somehow this trait has led me to be involved in many things over the years, which has greatly enriched both my career and my life.

It is through this interest in teaching and education that another very rewarding door was opened for me. Frank Swatek, who had been my advisor as an undergraduate and master's student, was Chairman of the Board of Education and Training (BET) at the American Society for Microbiology (ASM) during the early 1980s. I am not sure whether Frank figured I owed him one (which I did) or whether there were particular skills he was looking for, but I was recruited to be on the committee that put the workshop program together for ASM. Working with ASM turned out to be a rewarding experience—one I recommend to everyone in the profession. One thing led to another, and I found myself Chairman of the BET in 1991. This also got me a seat on the Council Policy Committee. I was and still am immensely impressed by the dedication and quality of ASM member volunteers and the contributions they unselfishly make to the society.

I had a clear goal of improving the breadth and quality of educational offerings in microbiology. The past had too much emphasis on curriculum and not nearly enough on service to the members. At about this time we recognized several constituencies related to microbiology that had not been on anyone's radar screen. Bob Krasner had done a great job of letting us know that there were many people who taught microbiology in a variety of institutions where teaching was their main focus. At the time ASM didn't offer much for these microbiologists. The attempt to remedy this situation began with sessions at the General Meeting focused on teaching. BET put together a Teaching Materials Exchange where first lesson plans and later software could be shared. The National Science Foundation sponsored a program to bring together researchers and teachers for a joint summer experience. The response to these activities was so positive that it was obvious that some additional effort was warranted, particularly for instructors from two-year institutions where over half the students taking microbiology get their first exposure to the science. (The number is more like 70% for minorities.) Jeff Sich, John Lammert, John Lennox, Phil Stukus, Kathy Jagger, and many others put together the first Undergraduate Education Conference. The focus of the conference was to define the content of the first course in microbiology. Out of that meeting evolved the core themes in microbiology, which have been used far beyond their initial intent. What started as a one-shot effort evolved into a highly successful regular meeting. The skills and dedication of this group of microbiologists are in every way the same as in research-based meetings.

Thanks to the efforts of many volunteers and the vision of ASM leadership over the six years I served as Chair, we were able to become much more responsive to member needs. We added a Minority Education Committee that generated a number of programs to involve additional minority students and faculty in the profession. We revitalized programs for precollege education and helped produce the wonderful Microbial Discovery Workshops for middle-school science teachers. We formed a Distance Education Committee to help bring the new electronic technology into our classrooms. Today ASM has an education program that is looked at as the model by other professional societies.

After six years, it was time to hand over the reins of BET. Somehow ASM has this way of not letting go of people, and I was sucked into another ASM project—the Microbial Literacy Collaborative. Cynthia Needham was directing the project for ASM and wanted help with generating a telecourse in microbiology using the images emerging from the *Intimate Strangers* prime-time PBS documentary. Fortunately, we had the core themes in microbiology so we knew what needed to be covered in a distance education course intended for nonmajors. I became part of the Collaborative's Executive Committee and Chair of the Science Advisors for the Telecourse. During the next year and a half, I was involved in things totally different from anything I have ever done—immensely creative and I believe worthwhile for the ASM and the profession. From the very beginning, we had a very talented team of microbiologists, all of whom, except me, teach the first course in microbiology.

New directions that allow me to use what I have learned about communications and teaching combined with my integrative research focus include spearheading the production of a telecourse on environmental sciences and conducting research on how digital video can be used in the classroom to both increase understanding among the students and change the way instructors teach. The tools are now becoming available that allow us to use a vast array of information resources in teaching and make the connection between what we do in the laboratory and what we teach in the classroom much tighter.

In those rare moments when there is time for retrospection, I often feel I have prospered from being in the right place at the right time, or at least having opportunities appear at fortunate moments. Although I don't put much credence in luck, I have been very lucky: lucky to have picked the most interesting and significant field of science to study; lucky to have gotten a great education that led to an interesting and productive career; lucky to have lived and worked with such an outstanding group of colleagues and friends; lucky to have two great children who are part of a close and loving family. Maybe even lucky to have been asked to write this piece. The only lasting words of wisdom I might timidly offer are to look around once in a while at how what you are doing relates to other parts of the world. This can lead to new and exciting aspects of your life.

FREDERIC K. PFAENDER received his bachelor's and master's degrees from California State University in Long Beach in 1966 and 1968, respectively. He received his doctoral degree in microbiology from Cornell University in 1971. In 1971, he joined the faculty of the University of North Carolina in the Department of Environmental Science and Engineering, where he is now a Professor. His research focuses on the environmental distribution and microbial degradation of organic compounds, especially molecules considered pollutants, and on microbial activities in salt marsh estuarine ecosystems, oceans, and groundwater.

The following papers are representative of his publications:

Guthrie, E. A., and F. K. Pfaender. 1998. Reduced pyrene bioavailability in microbially active soils. *Environ. Sci. Technol.* **32:**501–508.

Carmichael, L. M., R. F. Christman, and F. K. Pfaender. 1997. Desorption and mineralization kinetics of phenanthrene and chrysene in contaminated soils. *Environ. Sci. Technol.* **31:**126–132.

Dobbins, D. C., and F. K. Pfaender. 1988. Methodology for assessing respiration and cellular incorporation of radiolabeled substrates by soil microbial communities. *Microb. Ecol.* **15:**257–273.

Aelion, C. M., C. M. Swindoll, and F. K. Pfaender. 1987. Adaptation to and biodegradation of xenobiotic compounds by microbial communities from a pristine aquifer. *Appl. Environ. Microbiol.* **53:**2212–2217.

Pfaender, F. K., and G. W. Bartholomew. 1982. Measurement of aquatic biodegradation rates by determining heterotrophic uptake of radiolabeled pollutants. *Appl. Environ. Microbiol.* **44:**159–164.

RONALD M. ATLAS

Always Teaching, Always Learning

Looking back, I can see the path with all its twists and turns that has led me to my current position in microbiology. It has been a career path full of serendipity and surprises. My fascination with science and microorganisms began early. By the seventh grade, I was carrying out experimental investigations at home on the effects of electromagnetic radiation on plants and of plant hormones on microorganisms and entering projects in science fairs. After my junior year in high school, I spent a summer at Cornell University in a National Science Foundation program that allowed me to take two courses in microbiology—one general survey course and the other an experimental methods course that allowed us to carry out investigative studies. The lectures on microbial ecology by Martin Alexander must have had a major impact, as I remember them to this day. Mine, like all careers of microbiologists, is punctuated with mentors and memories.

Despite this early interest in science and microbiology, I had no intention of becoming a microbiologist. My thoughts were on medicine and saving humanity from disease. My images were of diseases like tuberculosis and polio, as those diseases were still prevalent in the neighborhood in New York City where I grew up. A career as a physician, not as a scientist, seemed the likely career path as I went off to college. In fact, although I majored in biology, the only course in microbiology that I took as an undergraduate was a seminar course taught by Edward Battley in which my one contribution was a paper on alcoholic fermentation followed by an evening of sampling a great variety of wines. Battley served as my undergraduate advisor and would later suggest that I explore graduate studies at Rutgers.

While my career path was aimed at medicine, my real interest at Stony Brook was learning about the world. It was the 1960s, filled with protests about everything, and I was part of that quest for a better world. I spent weekends in Greenwich Village with the poet Alan Ginsburg. I was at Woodstock. I wandered through Europe and experienced the diversity of humanity. I met my wife at a corner of the 1967 Montreal Expo. I stood at Nietzsche's Oxford and Divinity Street. And then I decided to go to graduate school and become a microbiologist.

At Rutgers I was assigned to the Department of Microbiology and Biochemistry in the Agriculture School, which later became Cook College. But all graduate students took the same introductory course—a year-long course taught by sixty faculty members in microbiology. The course covered the breadth of microbiology. Each lecture was specialized and in depth. Diverse, unconnected topics followed one another. We were left to our own discussions and our discussions with individual faculty to form a coherent picture of microbiology. I was fortunate to be guided by David Pramer, who placed me in the laboratory of Richard Bartha to do my research and who both then and now has guided by career. Mentoring is an essential part of career development.

Working in Bartha's laboratory was a strange mix of formality and friendship. Bartha was born in Hungary and educated at the University of Göttingen in Germany, where he worked in the laboratory of Hans Schlegel. He brought European formality to the laboratory at Rutgers. As students, we feared him and always respectfully addressed him as Dr. Bartha or Professor Bartha or even Herr Professor. But students also frequently gathered at his house for dinners, and my wife and I even spent a week camping with him and his family. It was not until I had successfully defended my Ph.D. dissertation and he had presented me the option of whether to continue the formality of the relationship that I first called him Richard.

During my graduate years, I learned a great deal about what was involved in being a microbiologist. There were the courses, but, more important, there was the laboratory. Bartha had a wonderful way of teaching students methods and then sitting back while the results were generated. Initially, I tried to work on two projects—one a physiologic project on the requirement for nickel by hydrogen-utilizing bacteria and the other an ecological project on the microbial utilization of petroleum hydrocarbons. I had little success with either, and Bartha was clearly concerned by my lack of progress. David Pramer would later describe how Bartha had asked the faculty to remove me from the program because I had been there for a year and had yet to publish a paper. Fortunately, I was given more time to develop as a scientist.

Bartha went off on a sabbatical at Woods Hole with Holger Jannasch, and I continued to muddle around the laboratory. I still didn't understand what it took to successfully carry out a scientific investigation that could withstand critical peer review. I decided to focus my efforts on the oil degradation project and specifically the investigation of the factors limiting petroleum biodegradation in the oceans. This was just over a year after the Torrey Canyon oil spill, and there was great public interest in the environment. My naiveté proved useful. I didn't know that there were questions that weren't to be asked because of scientific dogma. By taking the wrong path, I made new discoveries, and that inspired me to work harder. I began to work day and night, tied to the laboratory bench by a quest for discovery. Our Saint Bernard dog, Bernie, would lie outside the laboratory door, patiently waiting, perhaps even to save me if an experiment went awry. Many of the experiments were with flammable solvents, and more than one of Bartha's students had blown up the laboratory, on one occasion forcing Bartha to escape through the window and lower himself two stories using a rope. Fortunately, there were no injuries, and I was never responsible. My wife and many of the other graduate student spouses would gather each night in the department's conference room. Between experimental procedures I would join them, and we would eat, talk, and drink. Our social life developed around the university, and we still have many good friends from those days.

In retrospect, it's hard to understand why our work on oil biodegradation was so important. The fact that low nutrient concentrations in the oceans limit the rates of hydrocarbon biodegradation should have been obvious. That overcoming those limitations could speed up the removal of petroleum pollutants also should have been clear. But it wasn't, and my first scientific presentation before the American Chemical Society

Lecturing at the Chinese University in Hong Kong.

showing that petroleum biodegradation in the oceans is nutrient limited was so controversial that the meeting had to be adjourned and the next scheduled paper canceled. And who would have predicted that from our work, bioremediation would emerge as a major biotechnological solution for cleaning up the environment of many pollutants? Or that twenty years later, I would work with Exxon to apply what I had discovered as a graduate student to the bioremediation of the *Exxon Valdez* Alaskan oil spill? Although those outcomes were unpredicted in 1972, it was clear by the time I finished graduate studies that I was on the road to becoming a productive scientist. Bartha's patience was rewarded with ten publications from my graduate studies. Moreover, we had established a collaborative relationship that continues to this day.

After finishing graduate studies I took a postdoctoral position at the Jet Propulsion Laboratory of the California Institute of Technology in an Antarctic research program that was part of the NASA Mars *Viking* lander project. The idea was to use the Antarctic dry valleys as a test site for detection systems that would be sent to Mars. Unfortunately, several aircraft that were supposed to carry me to the Antarctic either crashed or developed mechanical problems, so I was left working in a freezer room in the laboratory in Pasadena. At lunch we would engage in great philosophical discussions asking: What is life? and how could we design a universal experimental system that would detect all life—any time and any place? One day I proposed that I take the system, which measured the conversion of carbon dioxide to organic matter (one of the few universal reactions of living systems), to the Arctic. The proposal was accepted, and I was off to the Naval Arctic Research Laboratory at Point Barrow, Alaska. There, besides testing the life-detection system that eventually would be sent to Mars, I renewed my studies on oil biodegradation for the Office of Naval Research.

I continued working in Alaska after moving to the University of Louisville. I would spend summers working with graduate students exploring the microbial population of tundra and coastal waters. We expanded our studies to work through the winter. We began diving under the ice—even when surface air temperatures were –50°C—to study the diversity of microbial communities and the abilities of indigenous microbial populations to degrade pollutants. Many of these studies used numerical taxonomy to characterize diverse microbial populations, requiring extensive laboratory and computer analyses. I found myself managing a laboratory of eighteen people with all the inherent personnel and fiscal problems. And it provided the opportunity to collaborate with Rita Colwell and others.

Although the research was going well and was well funded, I felt left out of the molecular biology revolution. My research program was still focused on biochemistry and ecology. So I encouraged some graduate students to begin studying the environmental fate of recombinant bacteria. The aim was to detect and contain genetically engineered microorganisms that might deliberately be released into the environment. We explored methods for containing genetically engineered microorganisms by using suicide vectors. Working together with Asim Bej, Mike Perlin, and Sorin Molin, we struggled to increase the sensitivity for detecting microorganisms in soil and water. We couldn't do better than about 10,000 bacteria per gram of soil. Then one of my graduate students, Robert Steffan, received a vial of enzyme and some suggestions on how it might help us. It turned out to be *taq* polymerase, and we were soon running polymerase chain reactions (PCRs). I have never asked where the enzyme came from. I was just thrilled that we could pioneer the environmental applications of PCR. Within a year, I could proclaim that we could detect a single genetically engineered microorganism in 100 grams of soil or 1 liter of water. Along with several other students and colleagues, we would use PCR to detect pathogens and indicator organisms in waters, including the bacterium *Legionella* and the protozoan *Giardia*. We even figured out how to use PCR for differentiating live from dead microorganisms. We used PCR to identify areas of significant health risks. Not only had I moved into the realm of molecular biology but also I had managed to join environmental and health-related research.

With the research successes came local, national, and international recognition. There were requests for presentations at scientific meetings, and my students and I found ourselves frequently traveling around the world. I was ill prepared for the travel demands, which grew to about one trip per week. There were also requests to serve on various committees. Juggling time became a major challenge. At one point, I found myself serving on twenty committees at the University of Louisville alone. Member of the faculty senate, head of the arts and sciences personnel committee, head of the biology department's graduate committee, and chair of the university's academic excellence committee. The time commitment was enormous and drew me away from teaching and research. Later, I became Associate Dean of the Arts and Sciences College and was even more removed from the aspects of academic life that I enjoyed. After three years in that administrative post, I tried to return to the laboratory. But I am afraid my submission to the dark side—as most academics view administrators—won, and I am now Dean of the Graduate School—even further removed from the science and excitement of discovery. My daily discoveries have more to do with finding out which meetings I need to attend and which ones I have already missed.

So I have turned to trying to fit the discoveries of others into science policy and the larger picture of integrated science and society. I continue to carry out extensive service at the national and international level. I will never forget how my hand trembled when I voted to approve the first human gene therapy experiments as a member of the National Institutes of Health Recombinant DNA Advisory Committee (RAC). That vote followed two years of discussions about safety, ethics, and science. Those debates often were heated, as I discovered when I failed to notice the CNN cameras capturing me clashing with a woman representing handicapped groups over whether medical researchers should try to find a cure for blindness. I argued that bringing our understanding of molecular biology to the treatment of disease represented a historic step that would better human health. Besides my service on the RAC and various other government boards, I spent ten years chairing the American Society for Microbiology (ASM) Public and Scientific Affairs Board Environment Committee and now chair